PHOTO ATLAS FOR BIOLOGY

PHOTO ATLAS

FOR BIOLOGY

James W. Perry
University of Wisconsin—Fox Valley

David Morton
Frostburg State University

Contributions to Animal Diversity by
WAYNE A. YODER
Frostburg State University

WADSWORTH PUBLISHING COMPANY

I(T)P® An International Thomson Publishing Company

Belmont • Albany • Bonn • Boston • Cincinnati • Detroit • London • Madrid • Melbourne
Mexico City • New York • Paris • San Francisco • Singapore • Tokyo • Toronto • Washington

Biology Editor	JACK CAREY
Assistant Editor	KRISTIN MILOTICH
Editorial Assistant	KERRI ABDINOOR
Production Editor	DEBORAH COGAN
Managing Designer	CAROLYN DEACY
Designer	CLOYCE WALL
Print Buyer	KAREN HUNT
Art Editor	KEVIN BERRY
Copy Editors	CHRIS EVERS & MARY ROYBAL
Cover Designer	GARY HEAD
Cover Photographs	JAMES W. PERRY, DAVID MORTON, & R. RAESLY
Composition and Art Preparation	STEVEN BOLINGER
Printer	VON HOFFMANN PRESS

FOR MORE INFORMATION, CONTACT:

Wadsworth Publishing Company
10 Davis Drive
Belmont, California 94002
USA

International Thomson Publishing Europe
Berkshire House 168-173
High Holborn
London, WC1V 7AA
England

Thomas Nelson Australia
102 Dodds Street
South Melbourne 3205
Victoria, Australia

Nelson Canada
1120 Birchmount Road
Scarborough, Ontario
Canada M1K 5G4

International Thomson Editores
Campos Eliseos 385, Piso 7
Col. Polanco
11560 México D.F. México

International Thomson Publishing GmbH
Königswinterer Strasse 418
53227 Bonn
Germany

International Thomson Publishing Asia
221 Henderson Road
#05-10 Henderson Building
Singapore 0315

International Thomson Publishing Japan
Hirakawacho Kyowa Building, 3F
2-2-1 Hirakawacho
Chiyoda-ku, Tokyo 102
Japan

10

ISBN 0-534-23556-5

THIS WORK IS DEDICATED TO THE MEMORY OF
DR. ROBERT DALE WARMBRODT

Contents*

*The terms used to describe the organisms here are general names for large assemblages of organisms sharing a number of characteristics. Most of these terms lack taxonomic status but are commonly used descriptors.

TAXONOMY CONTENTS

Taxonomic classification is a source of conjecture. Recently, the botanical category known as "division" has been replaced by "phylum," bringing into agreement plant and animal biologists. Beyond that major change, however, there is still a gulf of discord. Open ten biology books and you are sure to find nine different schemes. For example, many of the algae are being placed in the phylum Protista, and even here there are those who believe "Protista" should be "Protoctista." The classification used in this Photo Atlas is more traditional. We recognize that your instructors may have their own preferences

PREFACE

Among the natural sciences, perhaps none is as visually oriented as biology. This recognition has resulted in creation of photo atlases that precede this one. However, an evolutionary adaptation that allowed humans (and a few other species) to reach their remarkable level of success was color vision. Not only is color more pleasing to the eye, differences in reflective wavelengths create color contrast, allowing us to gain more information from an image than one that is gray scale. Our decades of experience with student learning led us to the unequivocal conclusion that to be most useful, a color atlas should be created, and thus the product before you.

Beyond the desire to create a color reference manual for students, there were two other pragmatic reasons for undertaking this project. First, many instructors like to create their own laboratory manuals, tailoring them to the needs of their students and having the copy reproduced in-house. While this technique satisfies many objections to using professionally published lab texts, the illustrations reproduce poorly, and are never in color. Consequently, this color photo atlas can usefully accompany the in-house productions. Second, the atlas allows students the opportunity to review specimens they have seen in the laboratory. Few educational institutions can accommodate students' desires for comprehensive, ongoing review of specimens, especially at the moment when a student has the time to do so.

We suggest that review is best accomplished by self-testing. For self-testing to be effective, students must be able to check their answers. Students can simply cover the labels within the atlas as review takes place, uncovering them for an immediate check.

In creating this atlas, we sifted through thousands of images in our collections, and decided which among them were most important for inclusion in an atlas that is useful not only for general biology students, but for those learning in introductory botany and zoology courses as well. We believe we have succeeded, basing our choices on our own teaching experiences in all three types of courses.

Any work becomes a collaborative process involving many individuals other than the authors. The folks at Wadsworth have been, as usual, outstanding assistants. We wish to credit specifically our Publisher, Jack Carey, who convinced the company that the cost of color production was a worthy undertaking. Assistant Editor Kristin Milotich helped keep the ball moving. And Production Editor Deborah Cogan deserves more recognition than space will allow. We'd also like to recognize the heroic efforts of Steven Bolinger, whose creative talent and technical acumen made production of these pages possible. Last but not least, we wish to thank John Limbach for his contributions to this work. An interactive CD-ROM and a laserdisc each containing over 4,700 microslide images is available from BIODISC, INC.; 6963 Easton Court; Sarasota, FL 34238.

We welcome comments and suggestions for improvement from both instructors and students. Consequently, we have included our addresses, telephone numbers, and electronic mail addresses. Please, let us know how we can help you.

James W. Perry   *David Morton*

ABOUT THE AUTHORS

JIM PERRY is the Campus Dean at The University of Wisconsin–Fox Valley, where he also teaches General Botany. His academic training is broad, including a B.S. in Zoology and Ph.D. in Botany and Plant Pathology, all from the University of Wisconsin–Madison. Prior to returning to Wisconsin, he was a faculty member at Frostburg State University, serving as the Chair of Biology, and teaching introductory biology courses as well as upper-level offerings in fungi, algae, the plant kingdom, and electron microscopy.

James W. Perry

Department of Biological Sciences
University of Wisconsin—Fox Valley
1478 Midway Road, PO Box 8002
Menasha, WI 54932-8002
phone: 414-832-2610
e-mail: jperry@uwcmail.uwc.edu

DAVE MORTON is Chair of the Biology Department at Frostburg State University. After earning a B.S. in Zoology and teaching junior high school, he attended Cornell University, where he received a Ph.D. with a major in Histology and minors in Physiology and Biochemistry. For more than twenty years he has taught numerous introductory, upper-level, and graduate biology courses at the college/university level. Some of his more interesting research publications describe aspects of iron and fluid balance in vampire bats.

David Morton

Department of Biology
Frostburg State University
Frostburg, MD 21532-1099
phone: 301-687-4355
e-mail: d_morton@fre.fsu.umd.edu

Abbreviations Used in Figure Legends

c.s. cross section

DIC differential interference contrast microscopy

l.s. longitudinal section

live photo taken from a living specimen

prep. slide photo taken from a prepared slide

sec. section

w.m. whole mount

= indicates a synonymous term

× magnification (as compared to actual specimen)

Note on magnifications: We chose to calculate magnifications of the images in this Photo Atlas as compared to that of the actual specimen, rather than indicating which microscope objective lens or magnification was used when the original photo was taken. For example, if the caption reads "1/2×" the image is one-half the size of the specimen. Likewise, an image listed as "500×" would be five hundred times the actual size of the specimen.

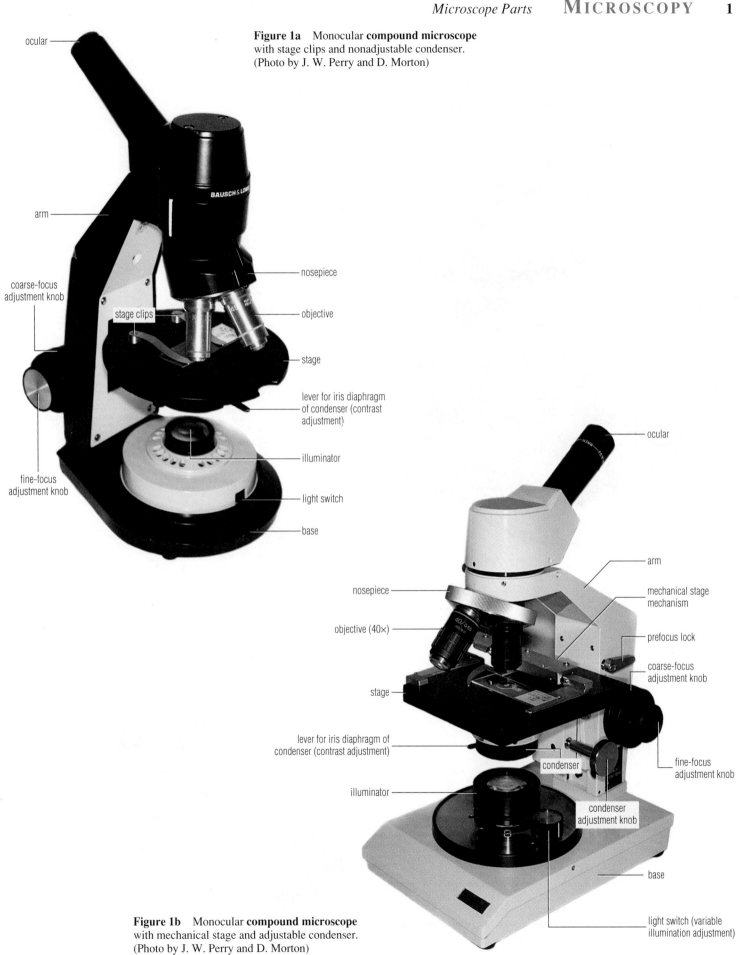

Figure 1a Monocular **compound microscope**
with stage clips and nonadjustable condenser.
(Photo by J. W. Perry and D. Morton)

ocular

arm

coarse-focus
adjustment knob

stage clips

fine-focus
adjustment knob

nosepiece

objective

stage

lever for iris diaphragm
of condenser (contrast
adjustment)

illuminator

light switch

base

ocular

nosepiece

objective (40×)

stage

lever for iris diaphragm of
condenser (contrast adjustment)

condenser

illuminator

arm

mechanical stage
mechanism

prefocus lock

coarse-focus
adjustment knob

fine-focus
adjustment knob

condenser
adjustment knob

base

light switch (variable
illumination adjustment)

Figure 1b Monocular **compound microscope**
with mechanical stage and adjustable condenser.
(Photo by J. W. Perry and D. Morton)

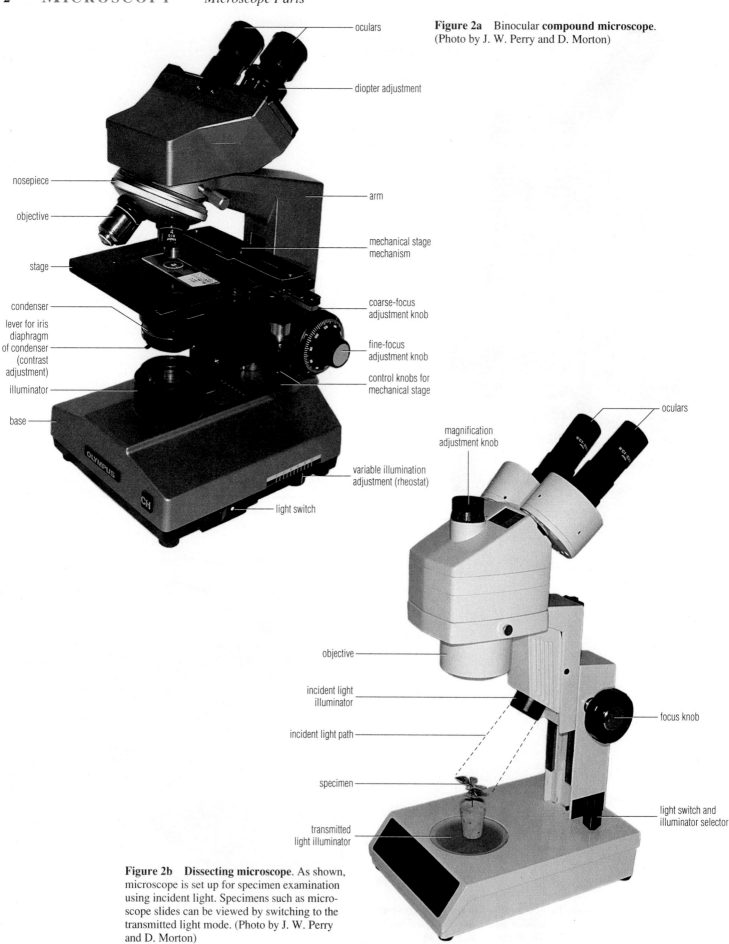

oculars

diopter adjustment

Figure 2a Binocular **compound microscope**.
(Photo by J. W. Perry and D. Morton)

nosepiece

objective

arm

mechanical stage
mechanism

stage

condenser

coarse-focus
adjustment knob

lever for iris
diaphragm
of condenser
(contrast
adjustment)

fine-focus
adjustment knob

illuminator

control knobs for
mechanical stage

base

oculars

variable illumination
adjustment (rheostat)

light switch

magnification
adjustment knob

objective

focus knob

incident light
illuminator

incident light path

specimen

light switch and
illuminator selector

transmitted
light illuminator

Figure 2b **Dissecting microscope**. As shown,
microscope is set up for specimen examination
using incident light. Specimens such as microscope slides can be viewed by switching to the
transmitted light mode. (Photo by J. W. Perry
and D. Morton)

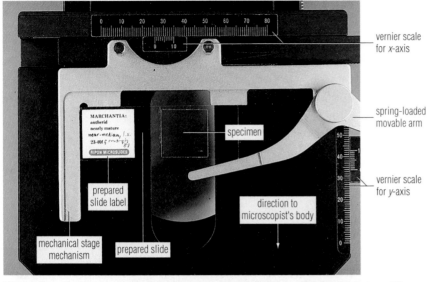

Figure 3d **Ocular** (eyepiece) removed from microscope. The markings on the ocular indicate this to be a "wide field" lens with a magnification of 10 (10×) and a focal length of 18.5 mm. (Photo by J. W. Perry)

Figure 3a Correct placement of prepared microscope slide on **mechanical stage**. The slide label is oriented so that it can be read by the microscopist. (Photo by J. W. Perry)

Figure 3b **Vernier scale** on mechanical stage of compound microscope. The correct reading is 19.6 mm. (Photo by J. W. Perry)

Figure 3e **10× objective**. Engravings indicate that the numerical aperture is 0.25, the mechanical tube length is 160 mm, and a coverslip 0.17 mm thick (No. 2 coverslip) must be used over the specimen for optimal resolution. (Photo by J. W. Perry)

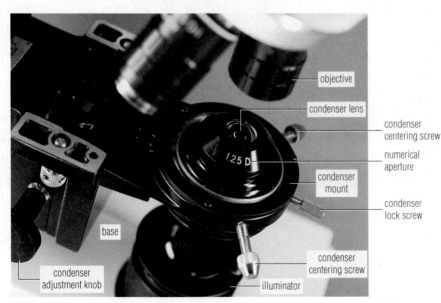

Figure 3c Adjustable **condenser** on a compound microscope (stage removed). The engraved number is the condenser lens's numerical aperture. (Photo by J. W. Perry)

Figure 3f **100× oil immersion objective**. Engravings indicate that the numerical aperture is 1.25, the mechanical tube length is 160 mm, and the objective is to be used with a slide having a coverslip 0.17 mm thick (No. 2 coverslip). The black ring indicates that the gap between the lens and the microscope slide's coverslip is to be filled with immersion oil. (Photo by J. W. Perry)

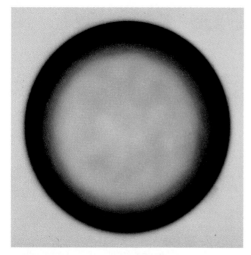

Figure 4a Free air bubble, a common "what's this?" in wet mounts (300×). (Photo by J. W. Perry)

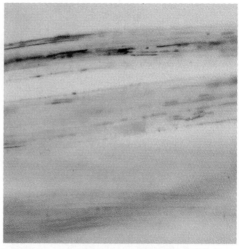

Figure 4c Low-contrast photo of unstained cotton fibers taken through microscope with iris diaphragm fully open (prep. slide, w.m., 100×). (Photo by J. W. Perry)

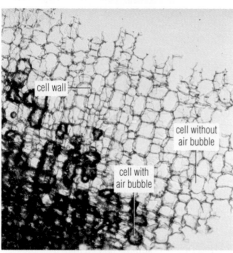

Figure 4b Cork cells. The darkened areas are cells filled with air (c.s., 100×). (Photo by J. W. Perry)

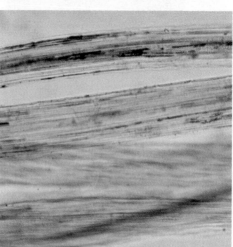

Figure 4d Increased-contrast photo of unstained cotton fibers taken through microscope with iris diaphragm fully closed (prep. slide, w.m., 100×). (Photo by J. W. Perry)

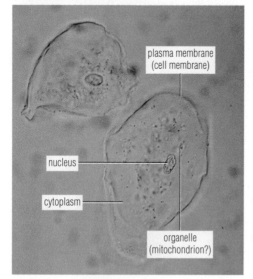

Figure 4e Unstained human cheek cells. Mottling in the background is due to a dirty lens (live, w.m., 450×). (Photo by J. W. Perry)

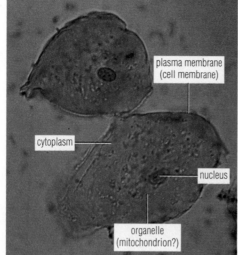

Figure 4f Human cheek cells stained with methylene blue. Specks in background are due to dirt on a lens (live, w.m., 500×). (Photo by J. W. Perry)

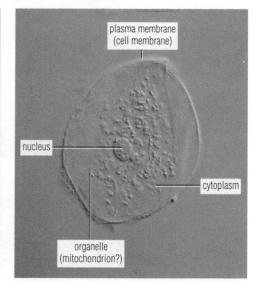

Figure 4g Unstained human cheek cells viewed with **differential interference contrast (DIC=Nomarski) microscopy**. This special technique produces a three-dimensional impression of the cell's organelles. Student microscopes are rarely equipped to produce such images (live, w.m., 400×). (Photo by J. W. Perry)

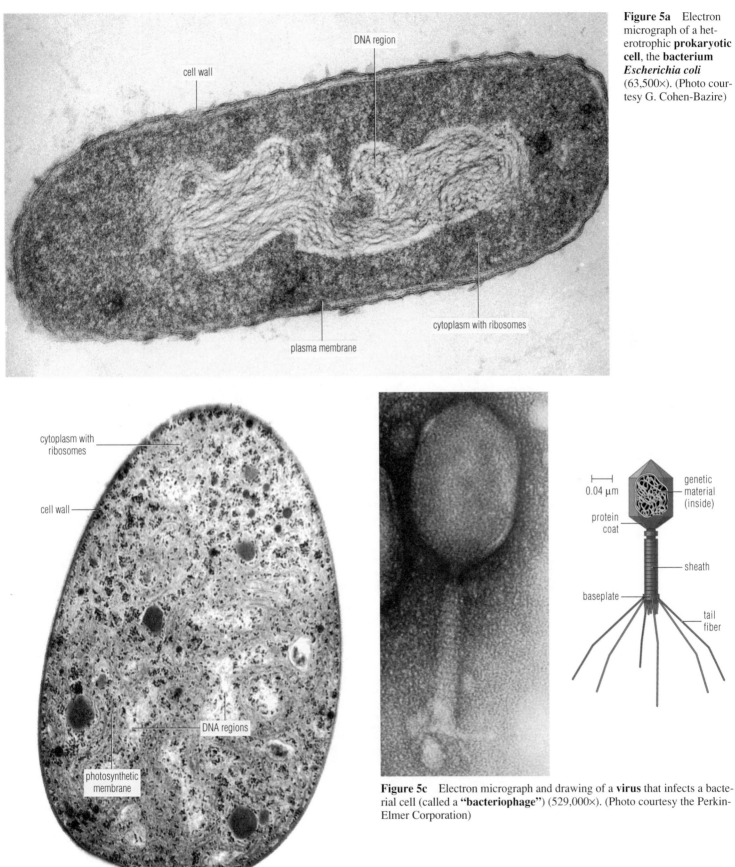

cell wall

DNA region

Figure 5a   Electron micrograph of a heterotrophic **prokaryotic cell**, the **bacterium *Escherichia coli*** (63,500×). (Photo courtesy G. Cohen-Bazire)

cytoplasm with ribosomes

plasma membrane

cytoplasm with ribosomes

cell wall

DNA regions

photosynthetic membrane

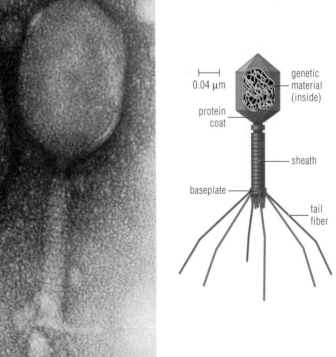

0.04 μm

protein coat

genetic material (inside)

sheath

baseplate

tail fiber

Figure 5c   Electron micrograph and drawing of a **virus** that infects a bacterial cell (called a **"bacteriophage"**) (529,000×). (Photo courtesy the Perkin-Elmer Corporation)

Figure 5b   Electron micrograph of a photosynthetic **prokaryotic cell**, the **cyanobacterium *Anabaena*** (14,000×). (Photo courtesy R. D. Warmbrodt)

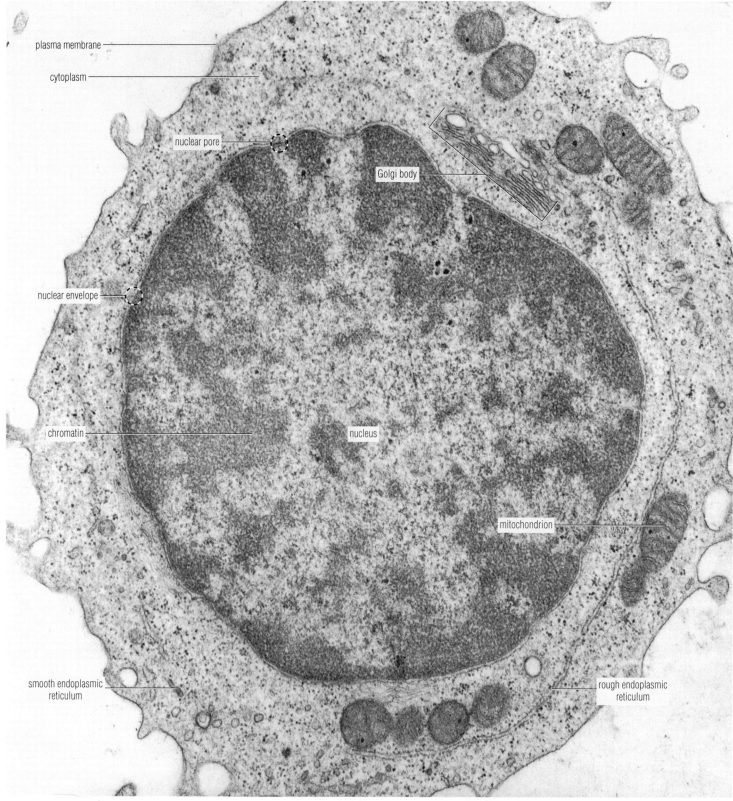

plasma membrane

cytoplasm

nuclear pore

Golgi body

nuclear envelope

chromatin

nucleus

mitochondrion

smooth endoplasmic
reticulum

rough endoplasmic
reticulum

Figure 6a Electron micrograph of an **animal cell** (3,000×). (Photo courtesy W. R. Hargreaves)

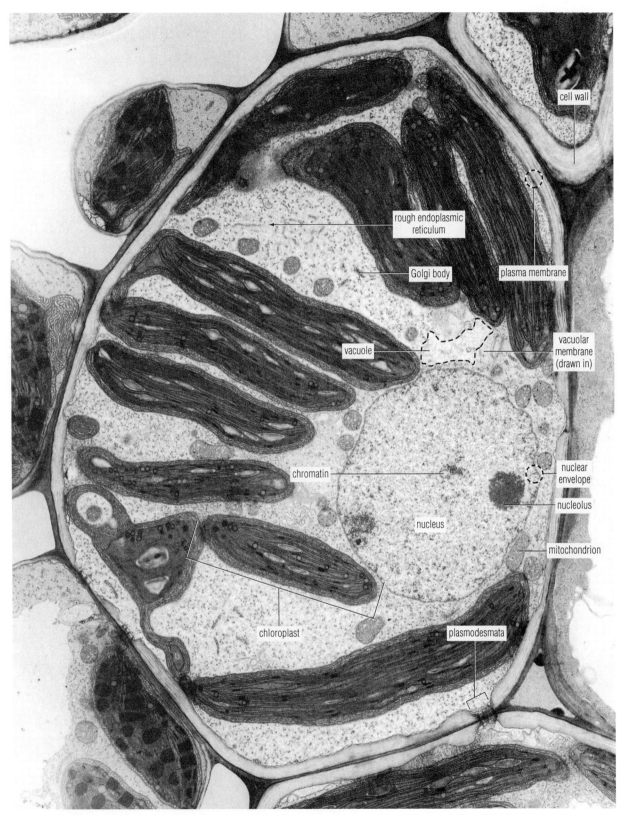

cell wall

rough endoplasmic
reticulum

Golgi body

plasma membrane

vacuole

vacuolar
membrane
(drawn in)

chromatin

nuclear
envelope

nucleolus

nucleus

mitochondrion

chloroplast

plasmodesmata

Figure 7a   Electron micrograph of a **plant cell** (c.s., 3,400×). (Photo courtesy R. F. Evert and M. A. Walsh)

Figure 8a Electron micrograph of **nucleus** from a rat's pancreas cell (13,000×). (Photo courtesy S. L. Wolfe)

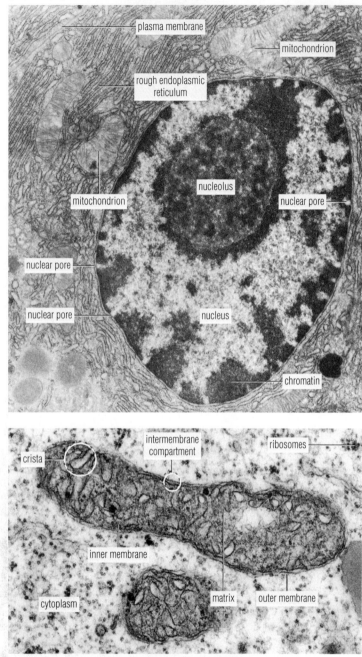

Figure 8b Electron micrograph of the **mitochondrion** and **rough endoplasmic reticulum** from a bat's pancreas cell (43,500×). (Photo courtesy K. R. Porter)

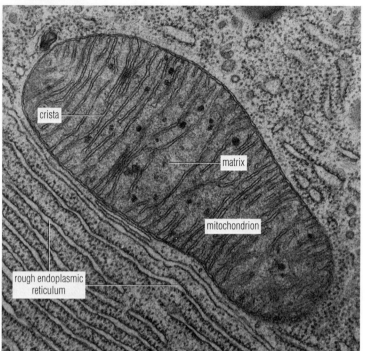

Figure 8c Electron micrograph of a **mitochondrion** from a plant (fern) cell. The intermembrane compartment is the region between the inner and outer membrane (13,500×). (Photo courtesy S. E. Eichhorn)

Figure 8d Electron micrograph of a **Golgi body**, sometimes also called a **dictyosome**, in a plant cell (17,000×). (Photo courtesy W. A. Jensen)

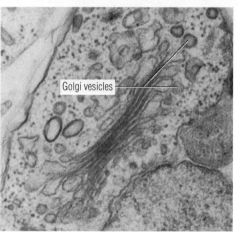

Figure 8e Electron micrograph of a plant cell's **chloroplast** (6,000×). Inset: a single granum (15,000×). (Photo courtesy R. R. Dute)

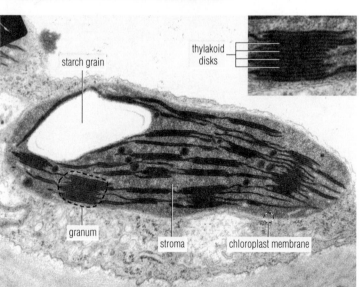

Figure 9a Electron micrograph of **microfilaments** (unlabeled arrows) in the endothelial cell of a human capillary. A few microtubules (mt) are also present (43,500×). (Photo courtesy K. G. Bensch, from *J. Ultrastr. Res.* 82:76 [1983])

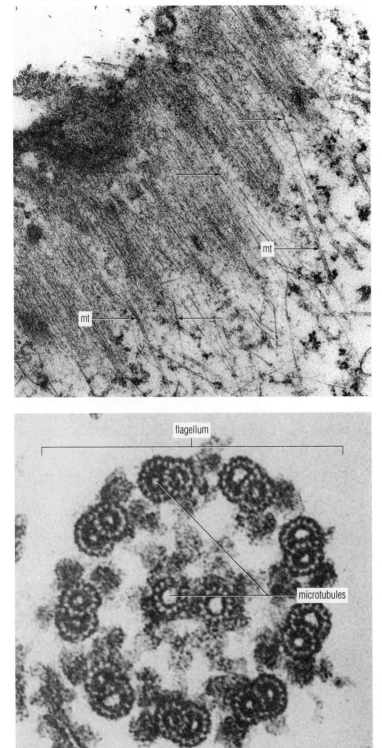

Figure 9b Electron micrograph of **microtubules** in longitudinal section running just beneath the cell wall of a plant cell (72,500×). (Photo courtesy E. H. Newcomb)

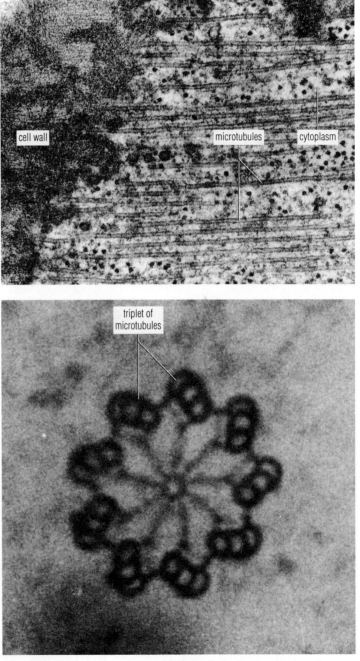

Figure 9c Electron micrograph of **microtubules** in cross section from a sea urchin flagellum (708,000×). (Photo courtesy K. Fujiwara, from *J. Cell Biol.* 59:267 [1963], by permission of the Rockefeller University Press)

Figure 9d Electron micrograph of **microtubules** in cross section of a centriole (232,000×). (Photo courtesy I. R. Gibbons)

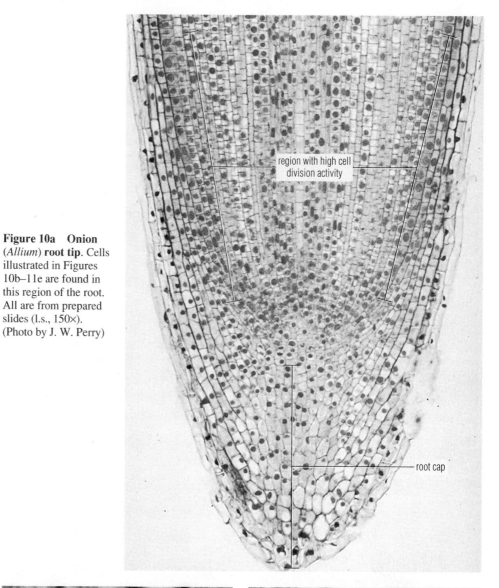

Figure 10a   Onion (*Allium*) **root tip**. Cells illustrated in Figures 10b–11e are found in this region of the root. All are from prepared slides (l.s., 150×). (Photo by J. W. Perry)

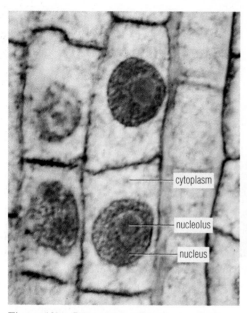

Figure 10b   Interphase cells prior to mitosis (l.s., 1400×). (Photo by J. W. Perry)

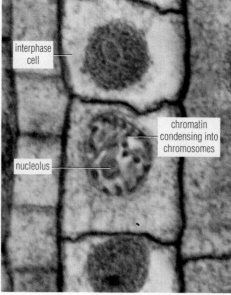

Figure 10c   Early prophase. Nuclear envelope intact (l.s., 1400×). (Photo by J. W. Perry)

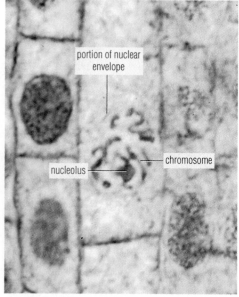

Figure 10d   Later prophase. Nuclear envelope disorganizing (l.s., 1400×). (Photo by J. W. Perry)

Figure 11a   Metaphase. Duplicated chromosomes (each consisting of two chromatids) at equatorial plane (l.s., 1400×). (Photo by J. W. Perry)

Figure 11b   Anaphase. Sister chromatids separating into unduplicated (daughter) chromosomes and moving toward opposite poles (l.s., 1400×). (Photo by J. W. Perry)

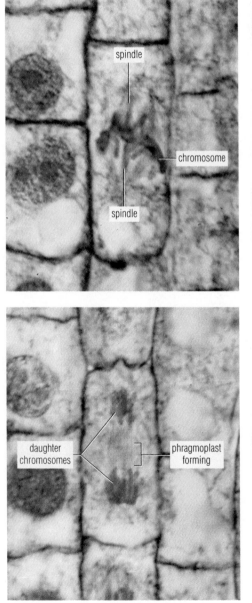

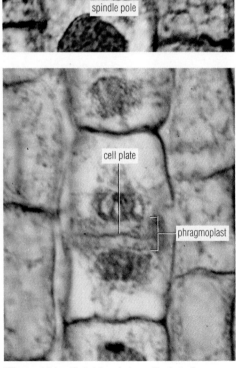

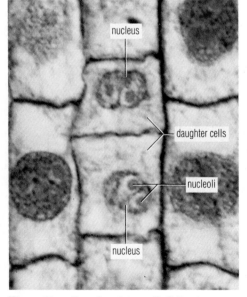

Figure 11c   Telophase. Unduplicated (daughter) chromosomes at poles (l.s., 1400×). (Photo by J. W. Perry)

Figure 11d   Cytokinesis by **cell plate formation**, a feature characteristic of most plant cells (l.s., 1400×). (Photo by J. W. Perry)

Figure 11e   Two **daughter cells** following cytokinesis. Note that both the nuclei and cytoplasmic volume in these two newly formed cells are smaller than in older neighboring cells (l.s., 1400×). (Photo by J. W. Perry)

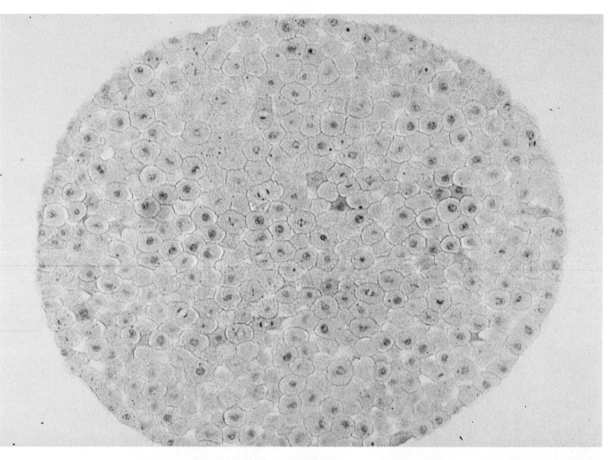

Figure 12a White-fish blastula. The blastula is a hollow ball of actively dividing cells produced early during embryonic development. Cells illustrated in Figures 12b–13e are found in this blastula (sec., 150×). (Photo by J. W. Perry)

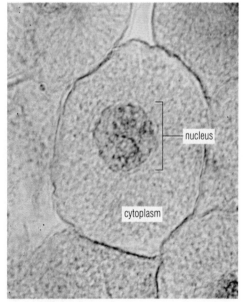

nucleus

cytoplasm

Figure 12b Interphase cell prior to mitosis (sec., 1000×). (Photo by J. W. Perry)

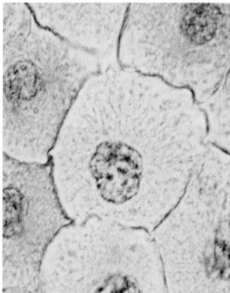

Figure 12c Prophase (sec., 1000×). (Photo by J. W. Perry)

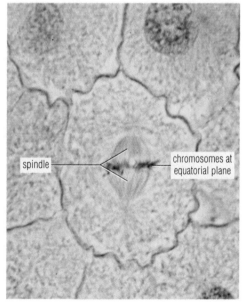

spindle

chromosomes at equatorial plane

Figure 12d Metaphase (sec., 1000×). (Photo by J. W. Perry)

Figure 13a **Early anaphase** (sec., 1000×). (Photo by J. W. Perry)

Figure 13b **Later anaphase** (sec., 1000×). (Photo by J. W. Perry)

Figure 13c **Telophase** (sec., 1000×). (Photo by J. W. Perry)

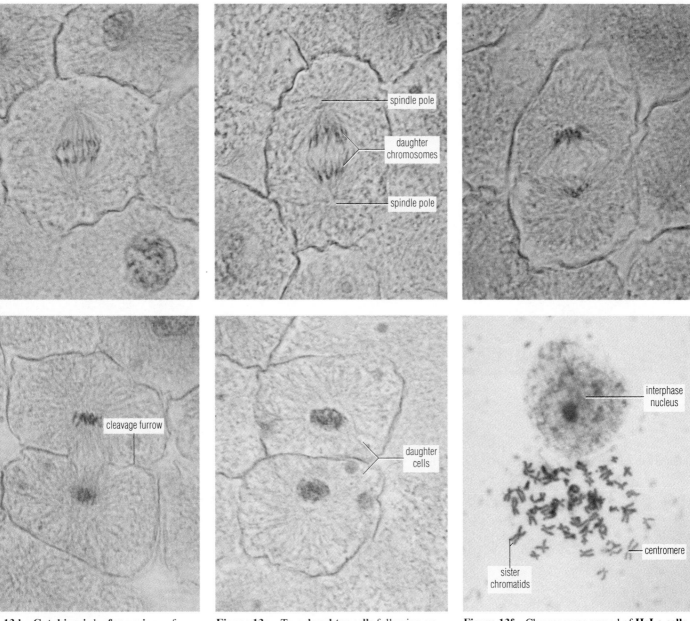

spindle pole

daughter chromosomes

spindle pole

cleavage furrow

daughter cells

interphase nucleus

centromere

sister chromatids

Figure 13d **Cytokinesis** by **furrowing**, a feature characteristic of most animal cells (sec., 1000×). (Photo by J. W. Perry)

Figure 13e Two **daughter cells** following cytokinesis (sec., 1000×). (Photo by J. W. Perry)

Figure 13f Chromosome spread of **HeLa cells**. Sister chromatids connected at their centromeres are clearly visible (w.m., 1000×). (Photo by J. W. Perry)

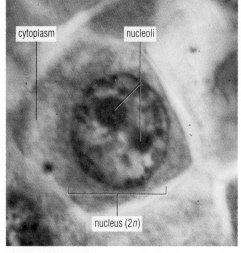

Figure 14a Diploid (2*n*) **interphase** cells in the anther of a lily (*Lilium*) flower (1600×). The nuclei in these cells undergo meiosis (Figures 14b–15g) to produce haploid sperm-producing pollen grains (all photos are sections). (Photo by J. W. Perry)

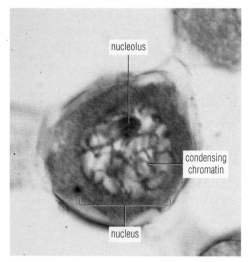

Figure 14b **Early prophase I** (sec., 1200×). (Photo by J. W. Perry)

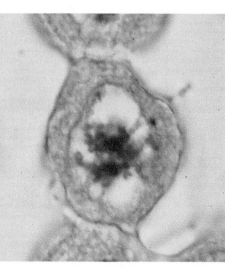

Figure 14c **Mid-prophase I** (sec., 1200×). (Photo by J. W. Perry)

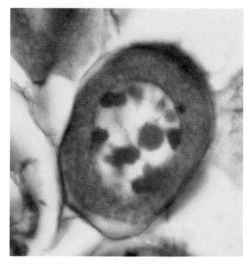

Figure 14d **Late prophase I** (sec., 1200×). (Photo by J. W. Perry)

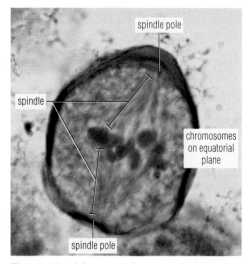

Figure 14e **Metaphase I** (sec., 1100×). (Photo by J. W. Perry)

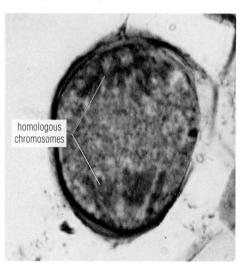

Figure 14f **Anaphase I**. The homologous chromosomes have separated (sec., 1300×). (Photo by J. W. Perry)

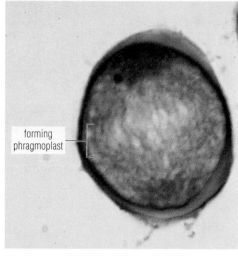

Figure 14g **Telophase I** (sec., 1300×). (Photo by J. W. Perry)

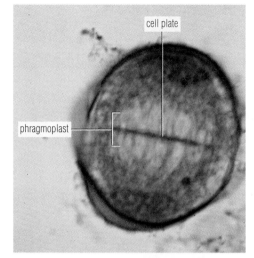

Figure 15a Meiosis I complete; **cytokinesis** by **cell plate formation** beginning (sec., 1200×). (Photo by J. W. Perry)

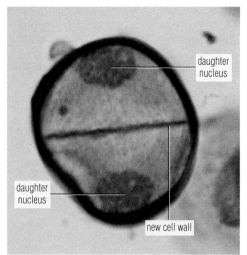

Figure 15b **Cytokinesis complete** (sec., 1300×). (Photo by J. W. Perry)

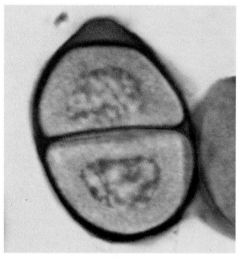

Figure 15c **Prophase II** (sec., 1300×). (Photo by J. W. Perry)

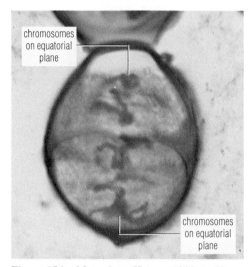

Figure 15d **Metaphase II** (sec., 1300×). (Photo by J. W. Perry)

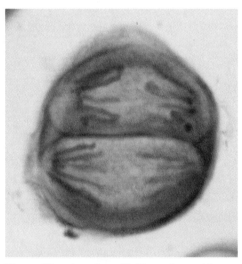

Figure 15e **Anaphase II**. Sister chromatids have seperated into unduplicated daughter chromosomes (sec., 1300×). (Photo by J. W. Perry)

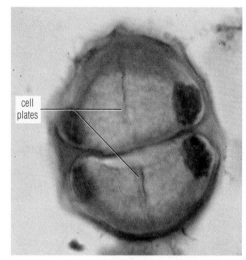

Figure 15f **Telophase II** and **cytokinesis** beginning in both cells (sec., 1300×). (Photo by J. W. Perry)

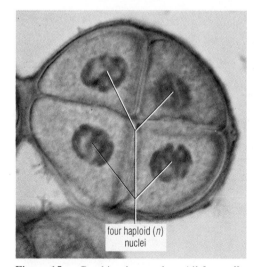

Figure 15g Cytokinesis complete. All four **cells are haploid (*n*)**. Following this stage, the four cells separate, each eventually forming sperm (sec., 1200×). (Photo by J. W. Perry)

Figure 16b Transmission electron micrograph of *Anabaena* (l.s., 14,000×). (Photo by Dr. R. D. Warmbrodt)

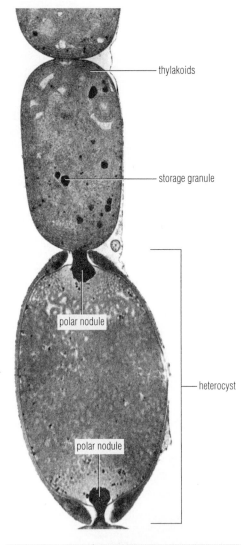

thylakoids

storage granule

polar nodule

heterocyst

polar nodule

Figure 16a
Anabaena azollae, a **cyanobacterium** that lives as a symbiont within the leaves of the water fern *Azolla*. Enlarged cells are heterocysts, functioning in nitrogen fixation (live, w.m. from a crushed *Azolla* leaf, 900×). (Photo by J. W. Perry)

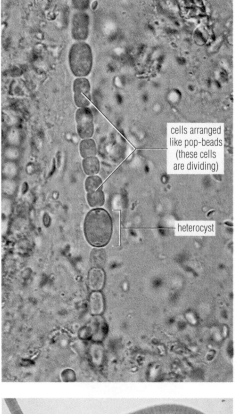

cells arranged like pop-beads (these cells are dividing)

heterocyst

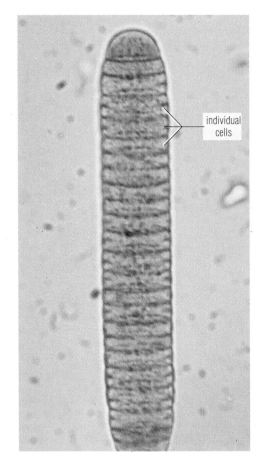

individual cells

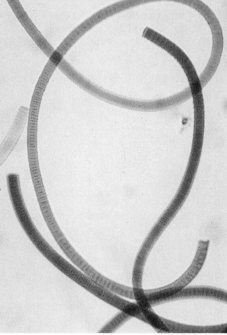

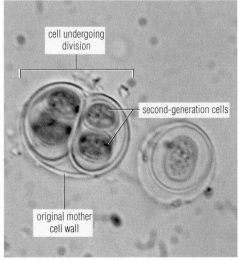

cell undergoing division

second-generation cells

original mother cell wall

Figure 16e *Gloeocapsa*, a **cyanobacterium** (live, w.m., 1400×). (Photo by J. W. Perry)

Figure 16c A short filament of the **cyanobacterium** *Oscillatoria* (live, w.m., 1300×). (Photo by J. W. Perry)

Figure 16d *Oscillatoria* (**cyanobacterium**) filaments (prep. slide, w.m., 350×). (Photo by J. W. Perry)

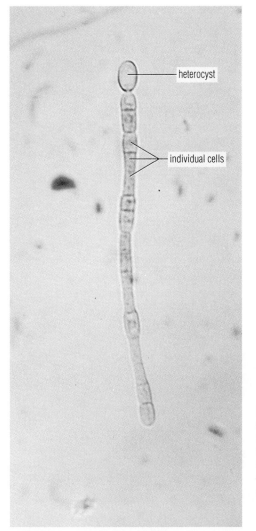

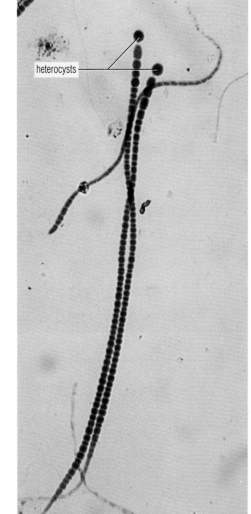

Figure 17a *Gloeotrichia*, a **cyanobacterium** (Phylum Gracilicutes) with a terminal heterocyst (live, w.m., 650×). (Photo by J. W. Perry)

Figure 17b *Gloeotrichia* **(cyanobacterium)** (prep. slide, w.m., 550×). (Photo by J. W. Perry)

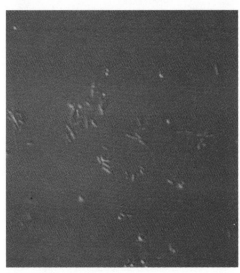

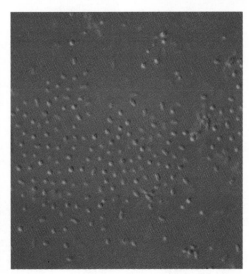

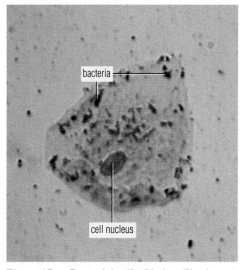

Figure 17c **Bacilli** (rod-shaped) **bacteria** (Phylum Firmicutes) from a sewage treatment sample (live, w.m., DIC microscopy, 1250×). (Photo by J. W. Perry)

Figure 17d **Cocci** (spherical) **bacteria** (Phylum Firmicutes) from a sewage treatment sample (live, w.m., DIC microscopy, 1250×). (Photo by J. W. Perry)

Figure 17e **Bacterial cells** (Phylum Firmicutes) on human cheek cell (prep. slide, w.m., 700×). (Photo by J. W. Perry)

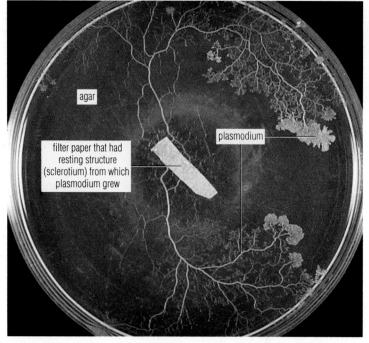

Figure 18a   Culture dish containing the **plasmodial slime mold** *Physarum polycephalum* (Phylum Gymnomycota) (1×). (Photo by J. W. Perry)

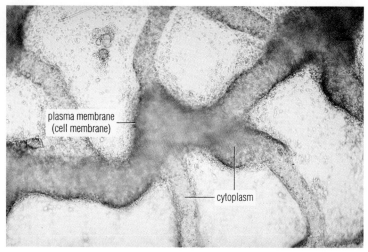

Figure 18b   Portion of **plasmodium** of *Physarum polycephalum* (live, 90×). (Photo by J. W. Perry)

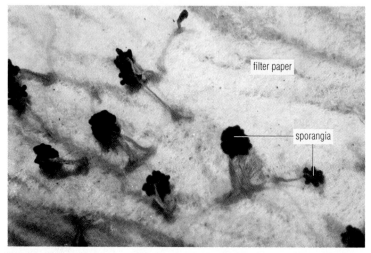

Figure 18c   Spore-containing **sporangia** of *Physarum polycephalum* (live, 1.5×). (Photo by J. W. Perry)

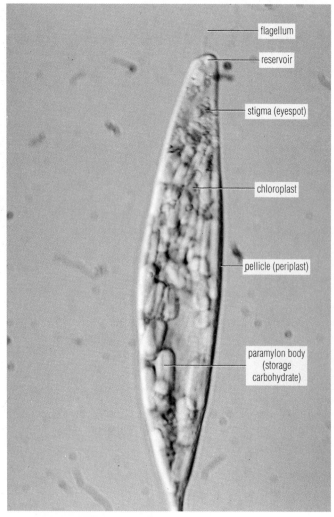

Figure 18d   *Euglena ascus* (Phylum Euglenophyta) (live, w.m., DIC microscopy, 1300×). (Photo by J. W. Perry)

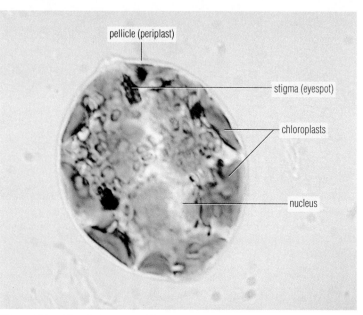

Figure 18e   *Trachelomonas* (Phylum Euglenophyta) (live, w.m., 1300×). (Photo by J. W. Perry)

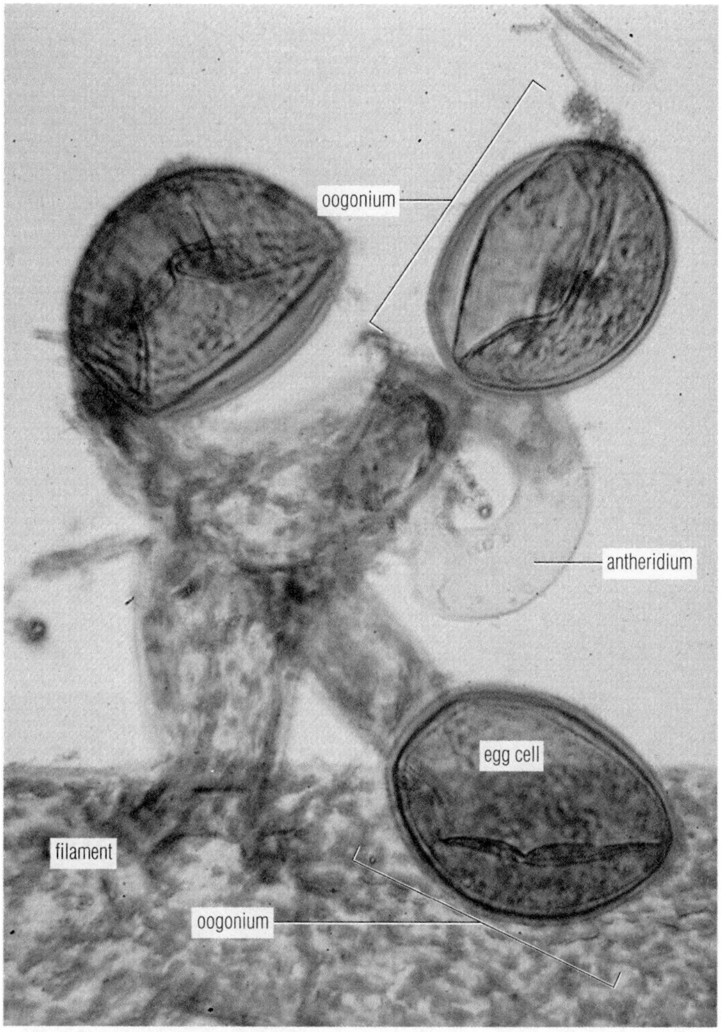

Figure 19a   Sex organs of *Vaucheria geminata* (prep. slide, w.m., 450×). (Photo by J. W. Perry)

oogonium

antheridium

egg cell

filament

oogonium

Figure 19b   *Vaucheria* zoosporangium (live, w.m., 300×). (Photo by J. W. Perry)

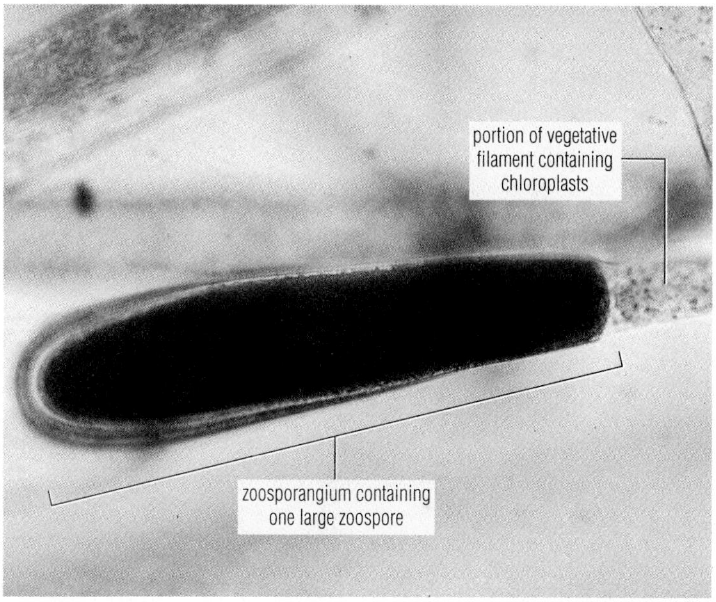

portion of vegetative filament containing chloroplasts

zoosporangium containing one large zoospore

Figure 19c   Single large **zoospore of *Vaucheria*** shortly after being shed from zoosporangium (live, w.m., 400×). (Photo by J. W. Perry)

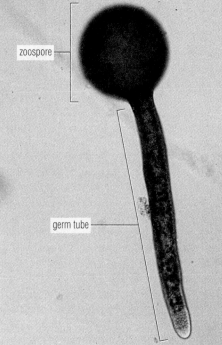

zoospore

germ tube

Figure 19d   Germinating **zoospore of *Vaucheria*** (live, w.m., 150×). (Photo by J. W. Perry)

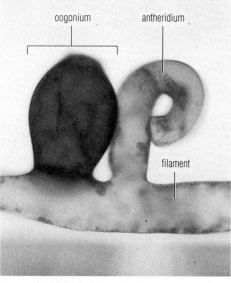

oogonium

antheridium

filament

Figure 19e   Sex organs of *Vaucheria sessilis* (prep. slide, w.m., 350×). (Photo by J. W. Perry)

**Figure 20a
Diatomaceous earth**
quarry near Quincy,
Washington. (Photo
courtesy Daniel F.
Williams)

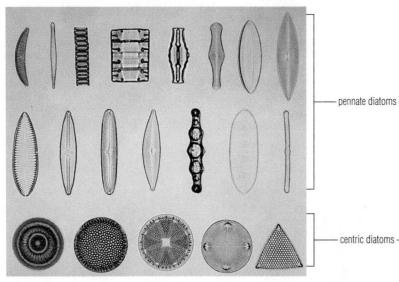

pennate diatoms

centric diatoms

Figure 20b   Mount of **diatoms** showing the variety of shapes
of different species (prep. slide, w.m., 100×). (Photo by J. W.
Perry)

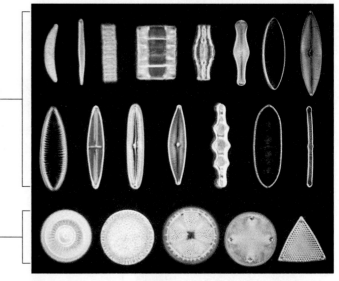

Figure 20c   Mount of **diatoms** (same as in Figure 20b), pho-
tographed with polarized light, which makes crystalline structures appear to "glow" against a dark background (100×). (Photo
by J. W. Perry)

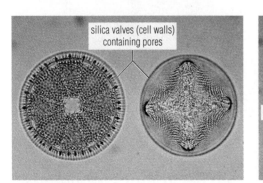

silica valves (cell walls)
containing pores

Figure 20d   Valve view of two **centric** (radially
symmetrical) **diatoms**, typical of those found in
marine (saltwater) environments (prep. slide,
w.m., 200×). (Photo by J. W. Perry)

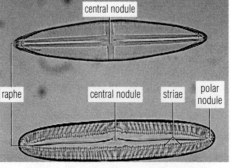

central nodule

raphe

central nodule

striae

polar
nodule

Figure 20e   Valve view of two **pennate** (bilaterally symmetrical) **diatoms**, typical of those found
in freshwater environments (prep. slide, w.m.,
250×). (Photo by J. W. Perry)

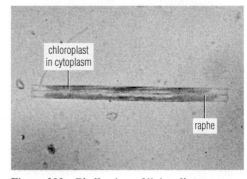

chloroplast
in cytoplasm

raphe

Figure 20f   **Girdle** view of living **diatom**
(w.m., 350×). (Photo by J. W. Perry)

Figure 21a The **dinoflagellate** *Ceratium* (prep. slide, w.m., DIC microscopy, 350×). (Photo by J. W. Perry)

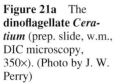

girdle (flagellar groove)

Figure 21b The **dinoflagellate** *Peridinium* (prep. slide, w.m., DIC microscopy, 1200×). (Photo by J. W. Perry)

girdle (flagellar groove)

girdle

nucleus

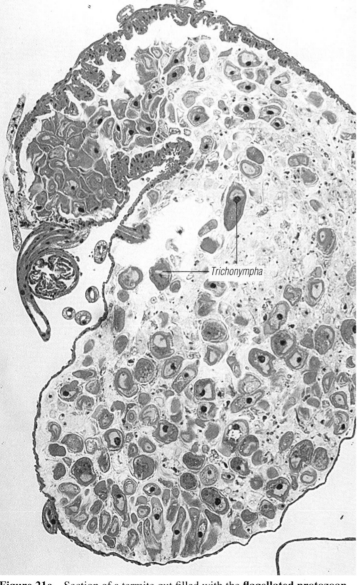

Trichonympha

Figure 21c Section of a termite gut filled with the **flagellated protozoan** *Trichonympha*, the organism that actually digests wood particles ingested by the termite (prep. slide, 150×). (Photo by J. W. Perry)

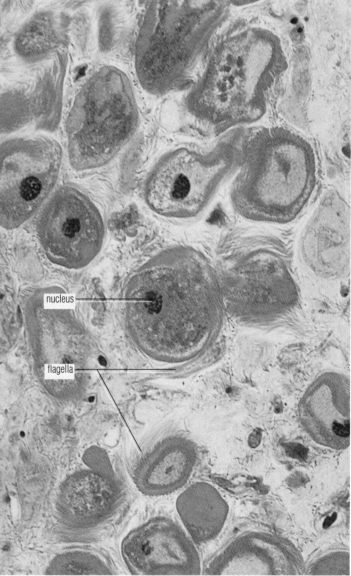

nucleus

flagella

Figure 21d The **flagellated protozoan** *Trichonympha* within the gut of a termite (prep. slide, sec., 600×). (Photo by J. W. Perry)

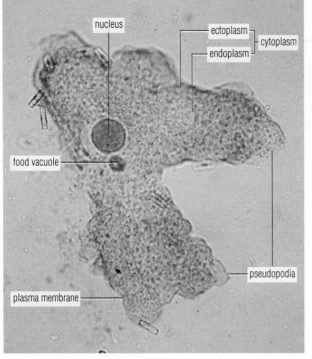

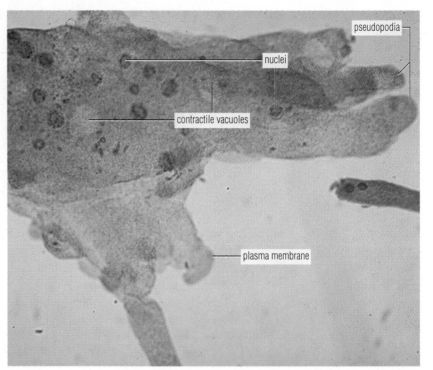

Figure 22a *Amoeba proteus* (prep. slide, w.m., 700×). (Photo by J. W. Perry)

Figure 22b The amoeboid protozoan *Chaos chaos* (prep. slide, w.m., 250×). (Photo by J. W. Perry)

Figure 22c Silica shell of a **marine amoeba**. Such shells are known as **radiolarians** (w.m., 300×). (Photo by J. W. Perry)

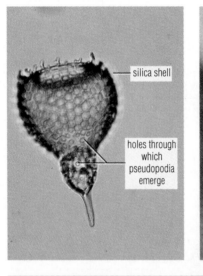

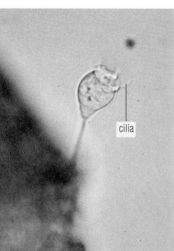

Figure 22d The ciliated protozoan *Vorticella microstoma* attached to a leaf (live, w.m., 600×). (Photo by J. W. Perry)

Figure 22e The ciliated protozoan *Paramecium* (prep. slide, w.m., 650×). (Photo by J. W. Perry)

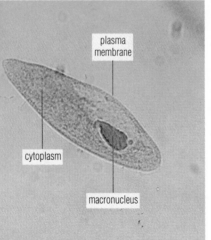

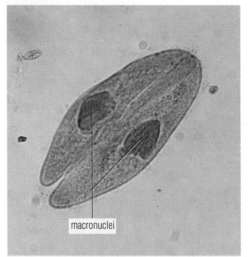

Figure 22f *Paramecium* undergoing conjugation (prep. slide, w.m., 600×). (Photo by J. W. Perry)

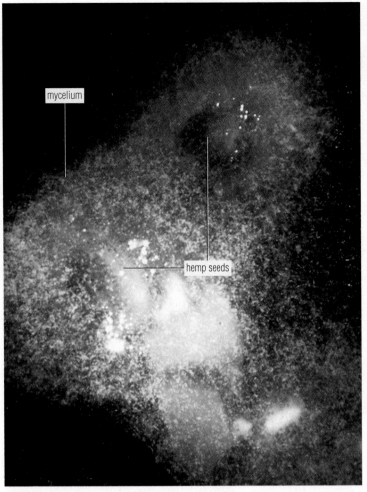

Figure 23a Mycelium of the **water mold** *Saprolegnia* growing on hemp seeds (live, 4.5×). (Photo by J. W. Perry)

Figure 23b *Saprolegnia* **zoosporangium** containing numerous zoospores (live, w.m., 450×). (Photo courtesy C. A. Taylor III)

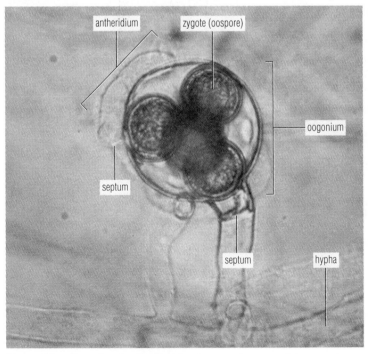

Figure 23c *Saprolegnia* **gametangia**, the oogonium containing zygotes (oospores) produced when sperm from the antheridium fertilized the egg cells (oospheres) (live, w.m., 550×). (Photo by J. W. Perry)

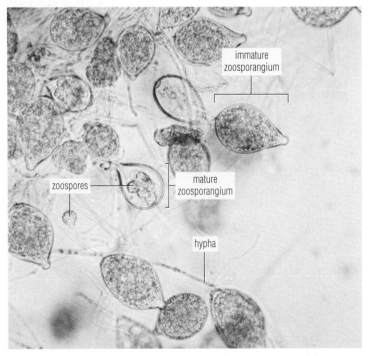

Figure 23d Pear-shaped **zoosporangia** containing zoospores of the water mold *Phytophthora* (live, w.m., 300×). (Photo by J. W. Perry)

Figure 24a
Mycelium of the **bread mold** *Rhizopus* growing on a slice of bread. The black dots are **sporangia** on the mycelium (live, 3×). (Photo by J. W. Perry)

mycelium with sporangia (black dots)

culture dish

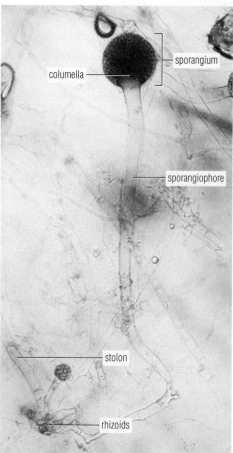

columella

sporangium

sporangiophore

stolon

rhizoids

Figure 24b **Sporangium** of *Rhizopus* atop sporangiophore (live, w.m., 80×). (Photo by J. W. Perry)

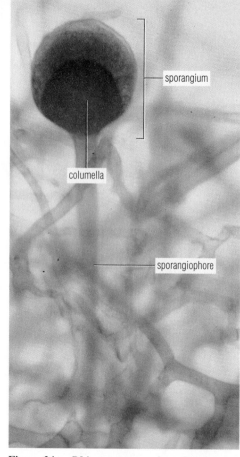

sporangium

columella

sporangiophore

Figure 24c *Rhizopus* **sporangium**. In this stained preparation, the columella is apparent (prep. slide, w.m., 150×). (Photo by J. W. Perry)

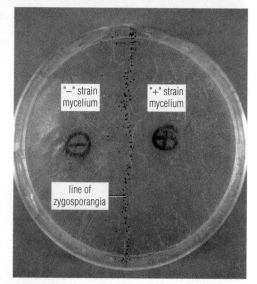

"−" strain mycelium

"+" strain mycelium

line of zygosporangia

Figure 24d Culture plate of a **bread mold** (***Phycomyces***) undergoing sexual reproduction. The vertical black line consists of **zygosporangia** formed as a result of fusion of gametangia produced by the "+" strain with gametangia produced by the "−" strain (live, 3/4×). (Photo by J. W. Perry)

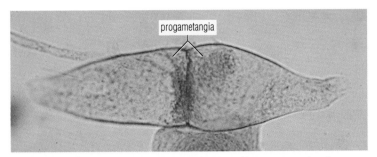

Figure 25a **Sexual reproduction in *Rhizopus*. Progametangia** of two strains have met (prep. slide, w.m., 300×). (Photo by J. W. Perry)

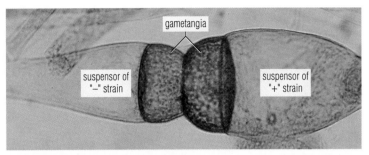

Figure 25b **Sexual reproduction in *Rhizopus*. Gametangia** have been produced from progametangia as a result of concentration of cytoplasm and subsequent formation of cell walls (prep. slide, w.m., 300×). (Photo by J. W. Perry)

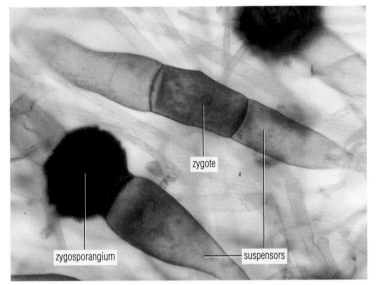

Figure 25c **Sexual reproduction in *Rhizopus*. Zygote** is formed by fusion of nuclei from respective gametangia after the walls at their tips break down and the cytoplasms fuse. Eventually, a thick wall forms around the zygote, producing the **zygosporangium** (prep. slide, w.m., 300×). (Photo by J. W. Perry)

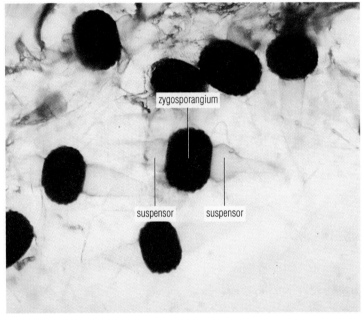

Figure 25d Living **zygosporangia of *Rhizopus*** (w.m., 100×). (Photo by J. W. Perry)

Figure 25e The **gun fungus *Pilobolus***, growing on horse dung. Sporangia at the tips of the enlarged bulbs (subsporangial vesicles) seen here may be shot up to 6 meters by a hydraulic mechanism (live, 1.2×). (Photo by J. W. Perry)

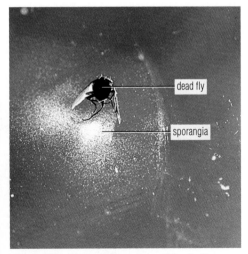

Figure 25f Halo of sporangia of the **fly fungus** *Entomophthora museae*. This fungus infects dead flies often found stuck on window glass (live, 1×). (Photo by J. W. Perry)

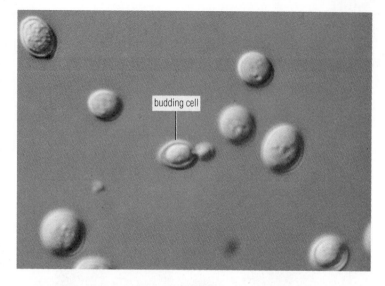

Figure 26a Cells of **bakers' yeast** *Saccharomyces*. One of these cells is reproducing asexually by **budding** (live, w.m., DIC microscopy, 1700×). (Photo by J. W. Perry)

Figure 26b Lilac leaf infected with the **powdery mildew** *Microsphaera*. The white mycelium covers the leaf and produces chains of **conidia** (live, 1×). (Photo by J. W. Perry)

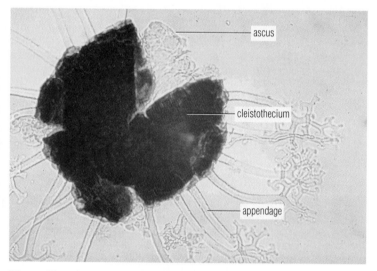

Figure 26c Ascus-containing **cleistothecium** of the **powdery mildew** *Microsphaera* (broken open). This appendaged sphere is the product of sexual reproduction and can be found as many "black dots" on leaves in late summer (live, w.m., 90×). (Photo by J. W. Perry)

Figure 26d Conidiophores bearing **conidia** of the **blue mold** *Eurotium*. This asexual stage is often named *Aspergillus* (prep. slide, w.m., 200×). (Photo courtesy Biodisc, Inc.)

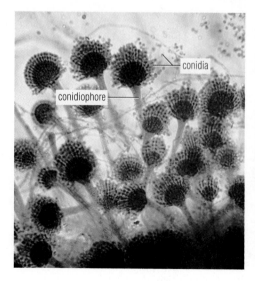

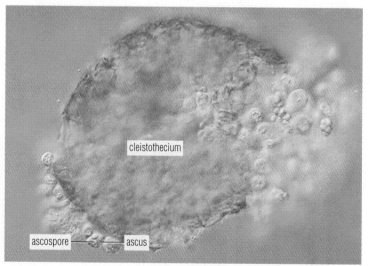

Figure 26e Sexually produced, ascus-containing **cleistothecium** (a type of ascocarp) of the **blue mold** *Eurotium chevalieri* that has been crushed to release the asci (live, w.m., DIC microscopy, 400×). (Photo by J. W. Perry)

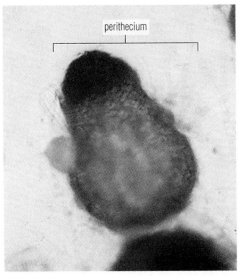

Figure 27a   Sexually produced, ascus-containing **perithecium** (a type of ascocarp) of the **sac fungus** *Sordaria* (live, w.m., 150×). (Photo by J. W. Perry)

Figure 27b   Sexually produced, ascus-containing **apothecia** (ascocarps) of the highly prized *Morchella*, a spring "mushroom" commonly called a **morel**. The asci line the ridges of this highly modified apothecium. These fungi are a true culinary delight (live, 2/5×). (Photo by J. W. Perry)

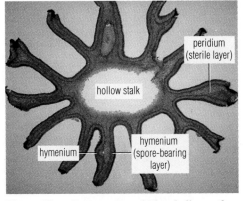

Figure 27c   Cross section of *Morchella* **apothecium** (ascocarp) with its two identifiable layers: the ascospore-containing hymenium and the sterile peridium (prep. slide, c.s., 16×). (Photo by J. W. Perry)

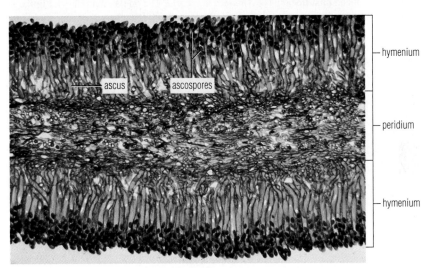

Figure 27d   **Asci** (singular, *ascus*) containing **ascospores** of *Morchella* (prep. slide, c.s., 300×). (Photo by J. W. Perry)

Figure 27e   **Apothecium** (ascocarp) of the **false morel,** *Gyromitra*. Some species are poisonous. They are distinguished from the true morels by the form of attachment of the "cap" to the stalk (live, 1/3×). (Photo by J. W. Perry)

Figure 27f   Sexually produced, ascus-containing **apothecium** of the **cup fungus** *Peziza*. The asci line the interior of the cup (live, 1/3×). (Photo by J. W. Perry)

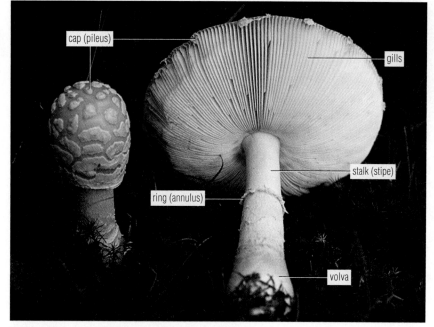

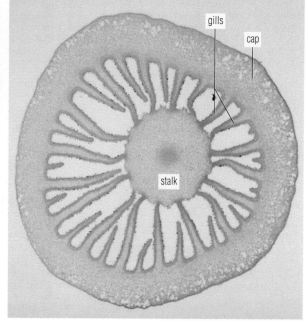

Figure 28a   Sexually produced **basidiocarps** of *Amanita muscaria*. The one on the left is younger, with its cap not yet fully expanded (live, 1×). (Photo by J. W. Perry)

Figure 28b   Cross section of the **cap** of the **club fungus** *Coprinus* (prep. slide, c.s., 20×). (Photo by J. W. Perry)

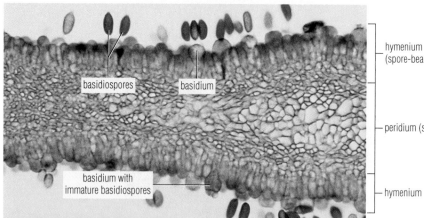

Figure 28c   *Coprinus* **gill with basidia-bearing basidiospores**. In this fungus, each basidium produces four basidiospores. The commercial mushroom commonly sold in grocery stores is *Agaricus bisporus*, which, as the specific epithet suggests, produces only two spores per basidium (prep. slide, c.s., 450×). (Photo by J. W. Perry)

Figure 28d   **Basidiocarp** of a **coral fungus** (live, 1/2×). (Photo by J. W. Perry)

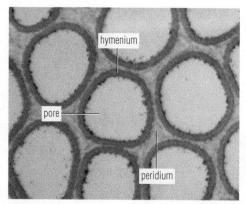

Figure 28e   **Shelf fungus** *Ganoderma*. This woody basidiocarp, called a **conk**, is an indication of mycelium growing within the tree (live, 1/10×). (Photo by J. W. Perry)

Figure 28f   Cross section of the lower surface of a **conk**. The basidia line the edges of the pores (prep. slide, c.s., 90×). (Photo by J. W. Perry)

Figure 28g   **Basidiocarp** of a **tooth fungus** (live, 1/3×). (Photo by J. W. Perry)

Figure 29a Mature **basidiocarps of puffballs**. The pore at the top of the puffballs allows the internally produced basidiospores to escape (live, 1/2×). (Photo by J. W. Perry)

Figure 29b Immature **basidiocarps of puffballs**. Before the basidiospores develop, the flesh is firm, as illustrated in the one that has been sliced open (live, 2/5×). (Photo by J. W. Perry)

Figure 29c   Basidiocarp of the stinkhorn fungus *Phallus*. The flesh of this fungus gives off an odor disagreeable to most humans but attractive to flies, which pick up the basidiospores, carrying them to new environments (live, 1×). (Photo by J. W. Perry)

Figure 29d   Basidiocarps of a bird's-nest fungus. The "eggs" contain the basidiospores (1/2×). (Photo by J. W. Perry)

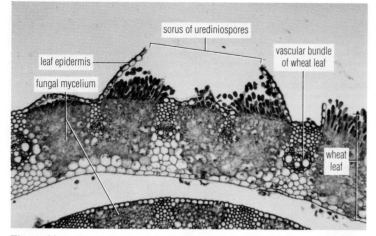

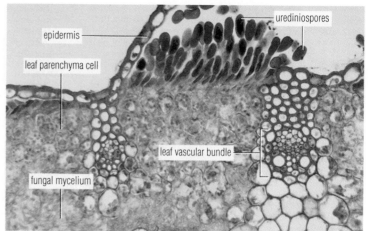

Figure 30a *Puccinia graminis*, the organism causing **wheat rust disease**. **Urediniospores** (= uredospores) have erupted from the surface of a wheat leaf. This is the **red** (uredial) **stage** of the disease, produced during North American summers (prep. slide, c.s., 20×). (Photo by J. W. Perry)

Figure 30b *Puccinia graminis* **urediniospores** growing from a mycelium-filled wheat leaf (prep. slide, c.s., 300×). (Photo by J. W. Perry)

Figure 30c *Puccinia graminis* **teliospores** erupting from the surface of a wheat leaf. This is the **black** (telial), overwintering **stage** of the disease (prep. slide, c.s., 300×). (Photo by J. W. Perry)

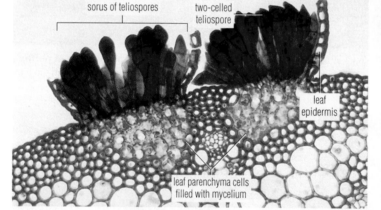

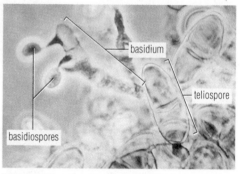

Figure 30d Germinating **teliospores** of the rust fungus *Gymnosporangium*. The teliospores have produced basidia on which basidiospores are borne (live, w.m., 300×). (Photo by J. W. Perry)

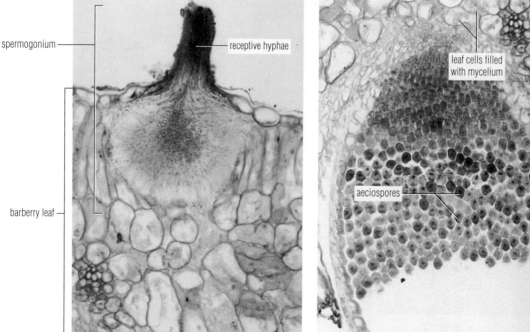

Figure 30e *Puccinia graminis* **spermogonium** (pycnium) on a barberry leaf, the alternate host of the pathogen (prep. slide, c.s., 300×). (Photo by J. W. Perry)

Figure 30f *Puccinia graminis* **aecium** containing **aeciospores** on a barberry leaf (prep. slide, c.s., 300×). (Photo by J. W. Perry)

Figure 31a   *Aspergillus* **conidiophores bearing conidia** (prep. slide, w.m., 300×). (Photo courtesy Biodisc, Inc.)

Figure 31b   *Aspergillus* **conidiophore and conidia** (live, w.m., 600×). (Photo by J. W. Perry)

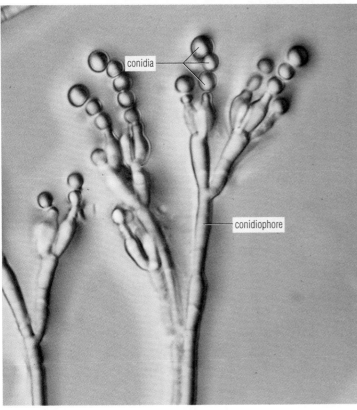

Figure 31c   *Penicillium* **conidiophore and conidia** (live, w.m., DIC microscopy, 550×). (Photo by J. W. Perry)

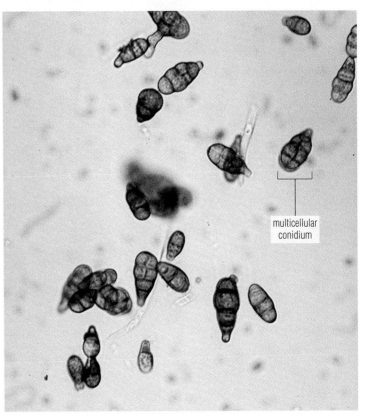

Figure 31d   *Alternaria* **conidia**. These airborne multicellular conidia cause an allergic reaction in many humans (live, w.m., 550×). (Photo by J. W. Perry)

Figure 32a Marine **red alga,** *Porphyra,* at low tide. This organism is sold dried in food stores as **nori** (live, 1/2×). (Photo by J. W. Perry)

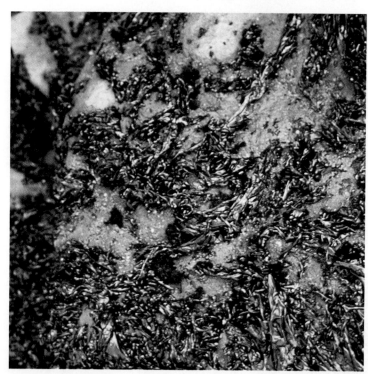

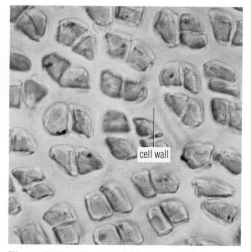

Figure 32b Microscopic appearance of *Porphyra* (w.m., dried specimen, 500×). (Photo by J. W. Perry)

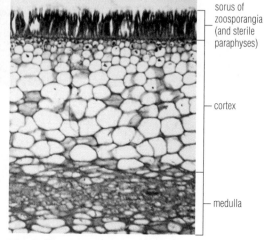

Figure 32e Portion of the **blade** of *Laminaria* (prep. slide, sec., 70×). (Photo by J. W. Perry)

Figure 32c **Red alga**, with coralline algae (live, 1/3×). (Photo by J. W. Perry)

Figure 32d *Laminaria,* a **marine brown alga** found along the northern coasts of the United States. Large marine brown algae are called **kelps** (live, 1/28×). (Photo by J. W. Perry)

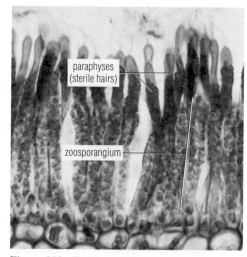

Figure 32f **Zoosporangia** of *Laminaria* (prep. slide, c.s., 300×). (Photo by J. W. Perry)

Figure 33a *Macrocystis*, a **kelp** found along the Pacific Coast of the United States (live, 1/20×). (Photo by J. W. Perry)

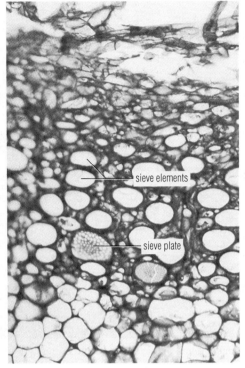

Figure 33b *Macrocystis* **stipe** in cross section. Specialized cells called "sieve elements," present in all vascularized land plants, have also evolved in some large brown algae to transport photosynthetic products (prep. slide, c.s., 150×). (Photo by J. W. Perry)

Figure 33c *Macrocystis* **stipe** in longitudinal section. See caption for Figure 33b (prep. slide, l.s., 100×). (Photo by J. W. Perry)

Figure 33d *Nereocystis*, a Pacific **kelp** with an air-filled vesicle called a pneumatocyst (live, 1/5×). (Photo by J. W. Perry)

Figure 34a   Colony of the Pacific **sea palm,** *Postelsia*. (Photo by J. W. Perry)

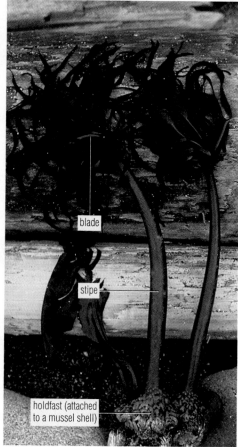

Figure 34b   *Postelsia,* the **sea palm** (live, 1/2×). (Photo by T. L. Carosella)

Figure 34c   *Fucus* (**rockweed**), a brown alga that grows along both coasts of the United States. The swollen receptacles at the ends of the branches contain the sex organs. "Dots" are openings to conceptacles shown in Figures 34d and 34e (live, 1/3×). (Photo by J. W. Perry)

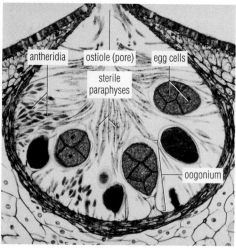

Figure 34e   *Fucus* **conceptacle** containing male antheridia and female oogonia. This is a monoecious species, having both male and female gametangia in the same conceptacle, and is found along the Atlantic Coast. Dioecious species (with male and female conceptacles found on separate plants) are found along the Pacific Coast (prep. slide, c.s., 90×). (Photo by J. W. Perry)

Figure 34d   *Fucus* **receptacle**. The cavities (conceptacles) are the location of the sex organs (prep. slide, c.s., 40×). (Photo by J. W. Perry)

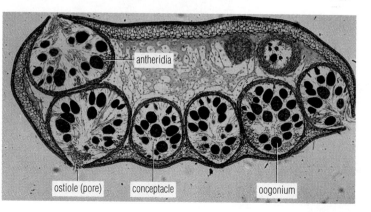

Figure 35a Transmission electron micrograph of the motile unicell *Chlamydomonas* (10,500×). (Photo courtesy H. Hoops)

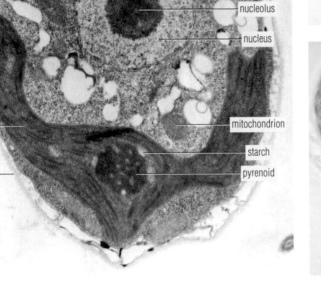
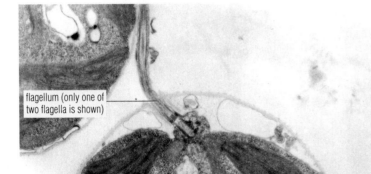

flagellum (only one of two flagella is shown)

chloroplast

nucleolus

nucleus

chloroplast

mitochondrion

cell wall

starch

pyrenoid

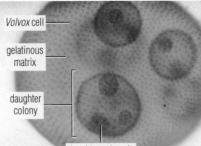

Volvox cell

gelatinous matrix

daughter colony

daughter colony in daughter colony

Figure 35b *Volvox* **colony** with asexually produced **daughter colonies** (autocolonies) (prep. slide, w.m., 10.4×). (Photo by J. W. Perry)

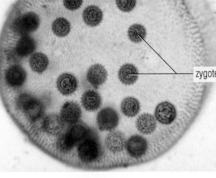

zygotes

Figure 35c *Volvox* **colony** with sexually produced **zygotes** (prep. slide, w.m., 150×). (Photo by J. W. Perry)

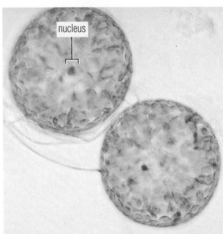

nucleus

Figure 35d Very large nonmotile unicells, *Eremosphaera*. These cells are so large they can be seen with the unaided eye (live, w.m., 250×). (Photo by J. W. Perry)

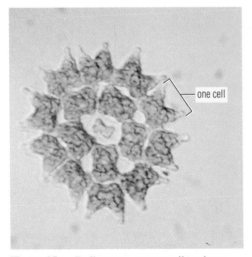

one cell

Figure 35e *Pediastrum*, a nonmotile colony (live, w.m., 650×). (Photo by J. W. Perry)

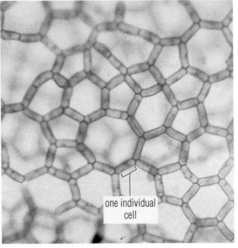

one individual cell

Figure 35f *Hydrodictyon*, the **water net**, a nonmotile colony (prep. slide, w.m., 250×). (Photo by J. W. Perry)

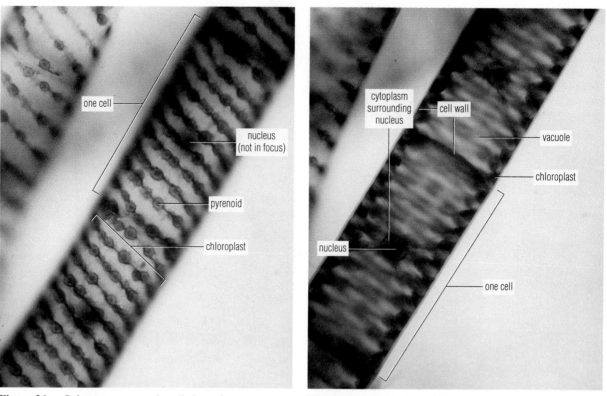

Figure 36a   *Spirogyra*, commonly called **pond scum**, a nonmotile filament. The focus is on the spiral **chloroplast** that has enlarged **pyrenoids**, centers for starch synthesis (prep. slide, w.m., 550×). (Photo by J. W. Perry)

Figure 36b   *Spirogyra*, with the focus on the **nucleus** (prep. slide, w.m., 550×). (Photo by J. W. Perry)

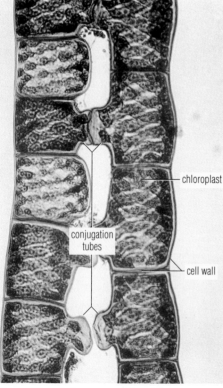

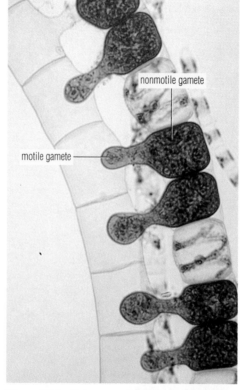

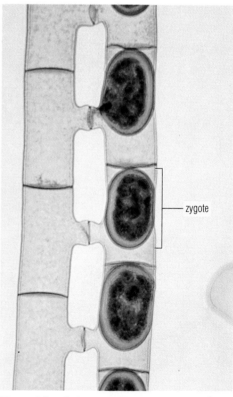

Figure 36c   *Spirogyra* undergoing sexual reproduction by **conjugation**. Conjugation tubes are meeting (prep. slide, w.m., 300×). (Photo courtesy Biodisc, Inc.)

Figure 36d   *Spirogyra* **conjugation**. The motile gamete from the left filament is migrating through the conjugation tube and fusing with the nonmotile gamete (prep. slide, w.m., 300×). (Photo courtesy Biodisc, Inc.)

Figure 36e   *Spirogyra* at the end of **conjugation**. Thick-walled **zygotes** have formed in one filament (prep. slide, w.m., 300×). (Photo courtesy Biodisc, Inc.)

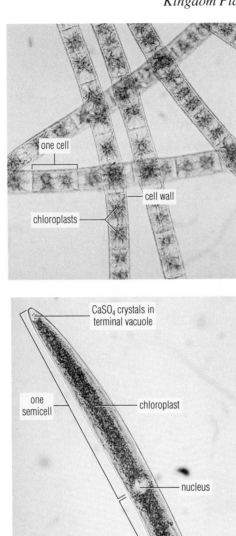

Figure 37a *Zygnema*, a nonmotile filament closely related to *Spirogyra* but with two star-shaped chloroplasts per cell (live, w.m., 250×). (Photo by J. W. Perry)

Figure 37d *Closterium*, a **desmid** (live, w.m., 250×). (Photo by J. W. Perry)

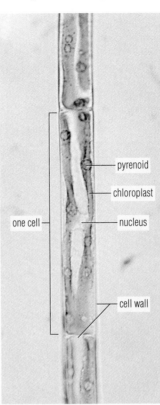

Figure 37b *Mougeotia*, a nonmotile filament with a ribbon-shaped chloroplast (live, w.m., 450×). (Photo by J. W. Perry)

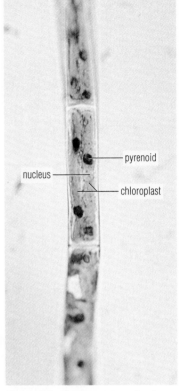

Figure 37c *Mougeotia* stained with potassium iodide (I_2KI) solution, which stains the starch-containing **pyrenoids** (live, w.m., 300×). (Photo by J. W. Perry)

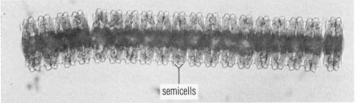

Figure 37f *Desmidium*. This **desmid** forms a chain, because the cells do not separate following cell division (live, w.m., 350×). (Photo by J. W. Perry)

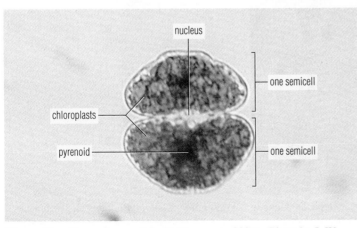

Figure 37e *Cosmarium*, a **desmid** (live, w.m., 800×). (Photo by J. W. Perry)

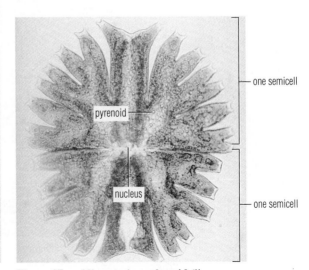

Figure 37g *Micrasterias*, a **desmid** (live, w.m., 250×). (Photo by J. W. Perry)

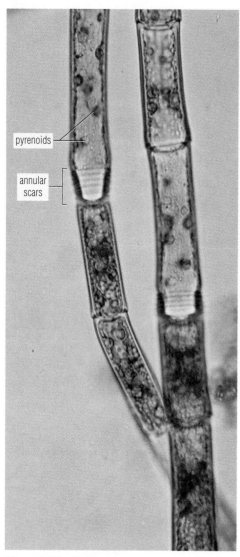

pyrenoids

annular scars

Figure 38a *Oedogonium*, a nonmotile filament. Annular scars are produced by successive cell divisions that occur within the filament (live, w.m., 550×). (Photo by J. W. Perry)

antheridia

oogonium

holdfast cell

Figure 38b *Oedogonium* filament that has gametangia and holdfast cell that anchors the filament to the substrate (prep. slide, w.m., 400×). (Photo courtesy Biodisc, Inc.)

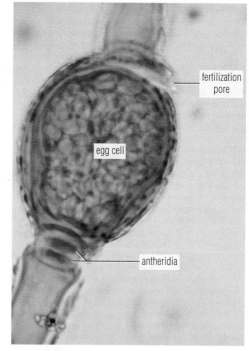

fertilization pore

egg cell

antheridia

Figure 38c *Oedogonium* gametangia. In this species, the sperm-containing **antheridia** are immediately adjacent to the **oogonium**. The pore through which the sperm enter the oogonium is visible (prep. slide, w.m., 800×). (Photo by J. W. Perry)

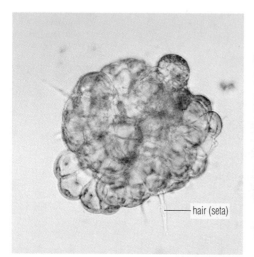

hair (seta)

Figure 38d *Coleochaete*, the alga believed to most closely resemble the ancestral alga that gave rise to land plants (live, w.m., 550×). (Photo by J. W. Perry)

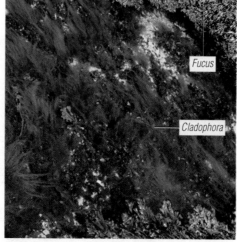

Fucus

Cladophora

Figure 38e *Cladophora* in a tidepool along the Atlantic Coast. Some *Cladophora* species are found in fresh water. (Photo by J. W. Perry)

Figure 38f *Cladophora* filament. The branching is characteristic of *Cladophora* (live, w.m., 100×). (Photo by J. W. Perry)

Figure 39a   *Codium,* **sea rope**, a marine siphonous green alga (live, 1/3×). (Photo by J. W. Perry)

Figure 39b   *Ulva,* **sea lettuce**, a marine tissuelike green alga (live, 2/5×). (Photo by J. W. Perry)

Figure 39c   *Chara* with sex organs (live, 3×). (Photo by J. W. Perry)

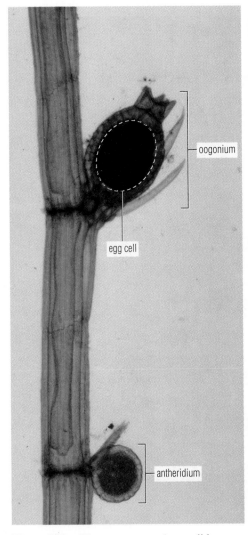

Figure 39d   *Chara* sex organs (prep. slide, w.m., 30×). (Photo by J. W. Perry)

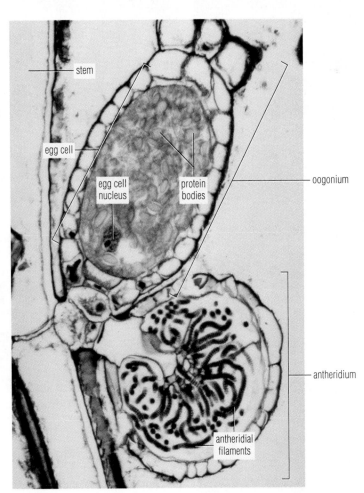

Figure 39e   *Chara* gametangia. In this individual, the gametangia are adjacent to one another. The single egg cell fills the oogonium (prep. slide, l.s., 100×). (Photo by J. W. Perry)

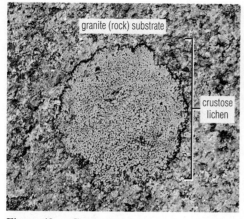

Figure 40a   Crustose lichen growing on rock (live, 1/3×). (Photo by J. W. Perry)

Figure 40b   Crustose lichen with ascospore-containing apothecia (live, 1/3×). (Photo by J. W. Perry)

Figure 40c   Foliose lichen (*Parmelia sulcata*) growing on the bark of a white birch tree (live, 1/3×). (Photo by J. W. Perry)

Figure 40d   Fruticose lichen (***Usnea***) on bald cypress tree (live, 1/3×). (Photo by J. W. Perry)

Figure 40f   Fruticose lichen. The cups at the tips of the branches are ascospore-containing apothecia (live, 1×). (Photo by J. W. Perry)

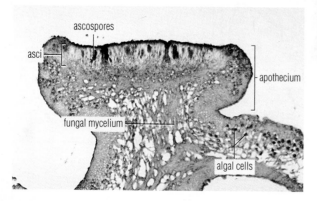

Figure 40e   Section of the **foliose lichen *Physcia*** with spore-bearing apothecium (ascocarp) (prep. slide, sec., 80×). (Photo by J. W. Perry)

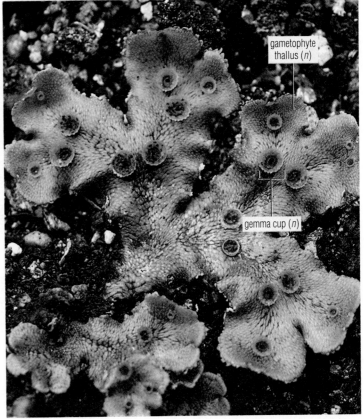

Figure 41a **Liverwort gametophyte thalli** (*Ricciocarpus*) (live, 1.2×). (Photo by J. W. Perry)

Figure 41b **Liverwort** (*Marchantia*) **thallus** with **gemma cups** containing gemmae, which can grow into new thalli (live, 2.4×). (Photo by J. W. Perry)

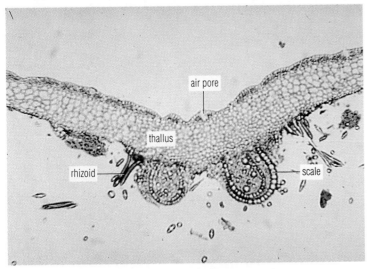

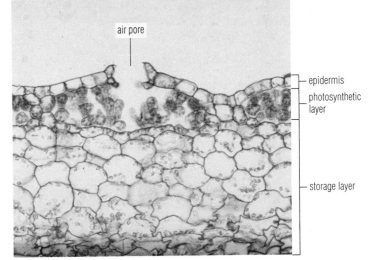

Figure 41c **Liverwort** (*Marchantia*) **thallus.** All structures are gametophytic and haploid (*n*) (prep. slide, c.s., 30×). (Photo by J. W. Perry)

Figure 41d **Liverwort** (*Marchantia*) **thallus** with air pore (prep. slide, c.s., 250×). (Photo by J. W. Perry)

Figure 42a Liverwort (*Marchantia*) **male gametophyte thalli** with **antheridiophores.** The umbrella-like platform at the top of the antheridiophore is sometimes called a "splash platform" (live, 1×). (Photo by J. W. Perry)

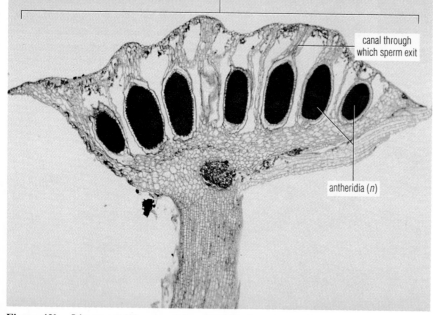

Figure 42b Liverwort (*Marchantia*) **antheridiophore** (prep. slide, l.s., 30×). (Photo by J. W. Perry)

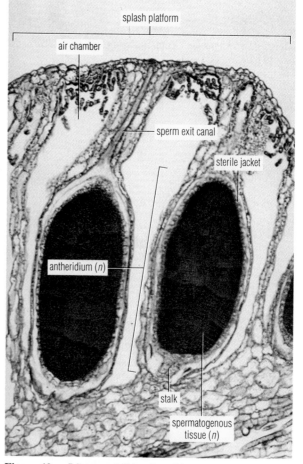

Figure 42c Liverwort (*Marchantia*) **antheridia** (prep. slide, l.s., 100×). (Photo by J. W. Perry)

Figure 43a Liverwort (*Marchantia*) **female gametophyte thalli** with **archegoniophores** (live, 1.3×). (Photo by J. W. Perry)

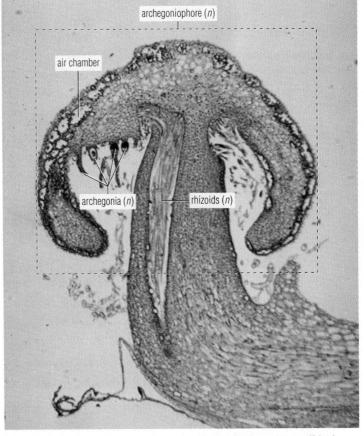

Figure 43b Liverwort (*Marchantia*) **archegoniophore** (prep. slide, l.s., 30×). (Photo by J. W. Perry)

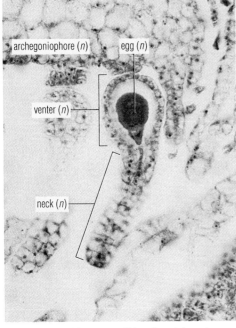

Figure 43c Liverwort (*Marchantia*) **archegonium** (prep. slide, l.s., 350×). (Photo by J. W. Perry)

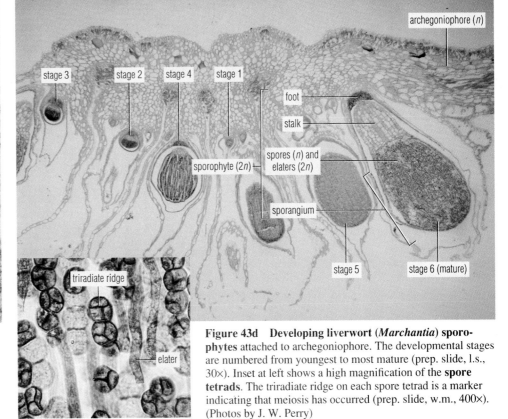

Figure 43d Developing liverwort (*Marchantia*) **sporophytes** attached to archegoniophore. The developmental stages are numbered from youngest to most mature (prep. slide, l.s., 30×). Inset at left shows a high magnification of the **spore tetrads**. The triradiate ridge on each spore tetrad is a marker indicating that meiosis has occurred (prep. slide, w.m., 400×). (Photos by J. W. Perry)

Figure 44a Hornwort (*Anthoceros*) **gametophyte thalli** with horn-like **sporophytes** (live, 2.5×). (Photo by J. W. Perry)

sporophytes (2*n*)

gametophyte thalli (*n*)

cyanobacteria on soil surface

sporangium (capsule) (2*n*)

sporophyte (2*n*)

gametophytes (*n*)

operculum (cap) (2*n*)

annulus

spores (*n*)

sporangium (capsule) (2*n*)

seta (stalk)

foot

Figure 44b *Sphagnum* **gametophytes** with **sporophytes** (live, 1×). (Photo courtesy Martyn Dibben)

Figure 44c *Sphagnum* **sporophyte** (prep. slide, l.s., 30×). (Photo by J. W. Perry)

Figure 45a Moss (*Polytrichum*) gametophytes. Antheridia are embedded in the cups atop the male gametophytes (live, 1/2×). (Photo by J. W. Perry)

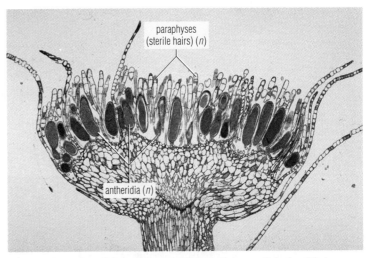

Figure 45b Moss (*Mnium*) antheridial head (prep. slide, l.s., 20×). (Photo by J. W. Perry)

Figure 45c Moss (*Mnium*) antheridia (prep. slide, l.s., 98×). (Photo by J. W. Perry)

Figure 45d Moss (*Mnium*) archegonial head (prep. slide, l.s., 30×). (Photo by J. W. Perry)

Figure 45e Moss (*Mnium*) archegonium (prep. slide, l.s., 100×). (Photo by J. W. Perry)

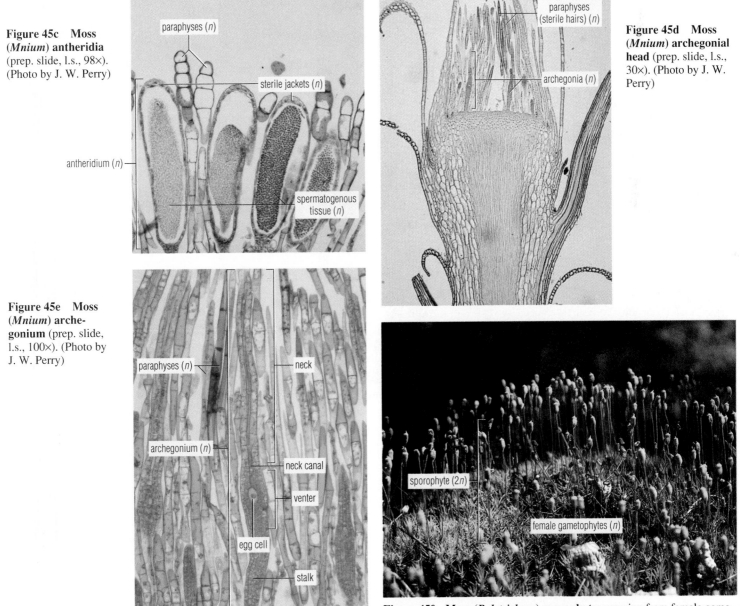

Figure 45f Moss (*Polytrichum*) sporophytes growing from female gametophytes (live, 4/5×). (Photo by J. W. Perry)

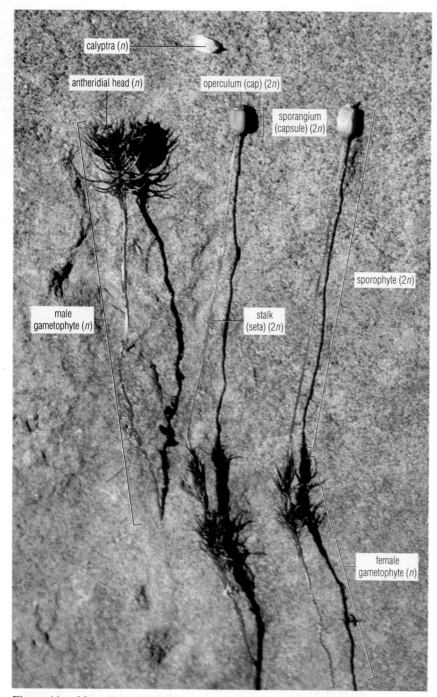

calyptra (*n*)

antheridial head (*n*)

operculum (cap) (2*n*)

sporangium (capsule) (2*n*)

sporophyte (2*n*)

male gametophyte (*n*)

stalk (seta) (2*n*)

female gametophyte (*n*)

Figure 46a   Moss (*Polytrichum*) male gametophyte, and female gametophytes with attached **sporophytes.** The **calyptra** has been removed from the leftmost sporangium (live, 1.4×). (Photo by J. W. Perry)

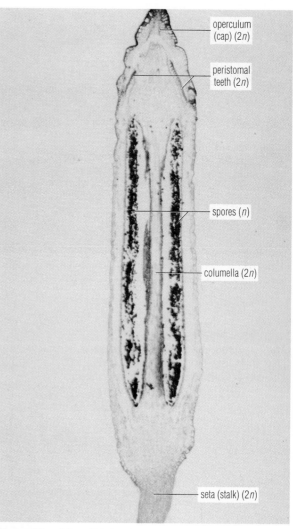

operculum (cap) (2*n*)

peristomal teeth (2*n*)

spores (*n*)

columella (2*n*)

seta (stalk) (2*n*)

Figure 46b   Moss (*Mnium*) sporangium (prep. slide, l.s., 8×). (Photo by J. W. Perry)

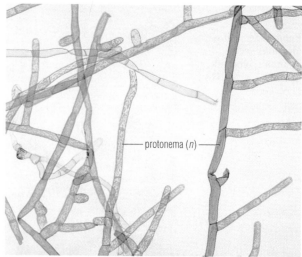

protonema (*n*)

Figure 46c   Moss (*Mnium*) protonema that has been grown from moss spores (prep. slide, w.m., 100×). (Photo by J. W. Perry)

Figure 47a   Whisk fern (*Psilotum nudum*) sporophyte. Note dichotomous branching of this aerial stem (live, 1×). (Photo by J. W. Perry)

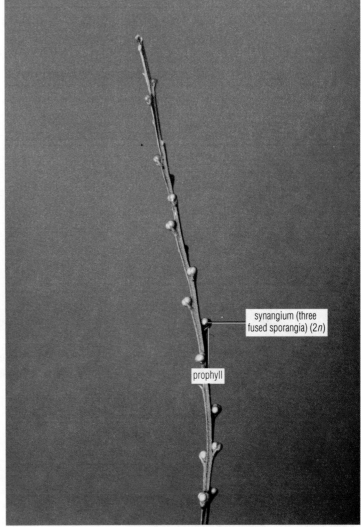

Figure 47b   Whisk fern (*Psilotum nudum*) aerial shoot with **sporangia** (live, 1.2×). (Photo by J. W. Perry)

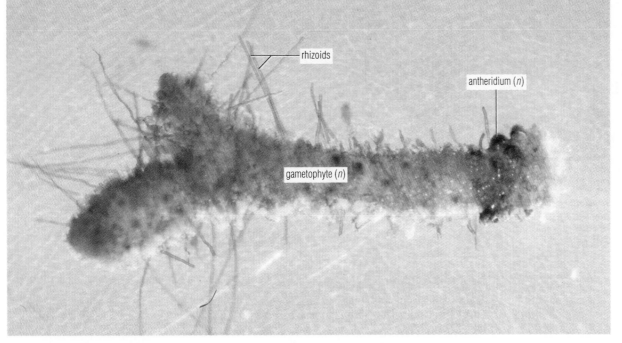

Figure 47c   Whisk fern (*Psilotum nudum*) gametophyte (live, 30×). (Photo by J. W. Perry)

Figure 48b Clubmoss (*Lycopodium obscurum*) **sporophyte.** The sporangia are aggregated in **cones (strobili)** at the top of the plant (live, 4/5×). (Photo by J. W. Perry)

Figure 48c Clubmoss, *Lycopodium lagopus* (*complanatum*), **sporophyte.** The cones are elevated on their own specialized branch in this species (live, 4/5×). (Photo by J. W. Perry)

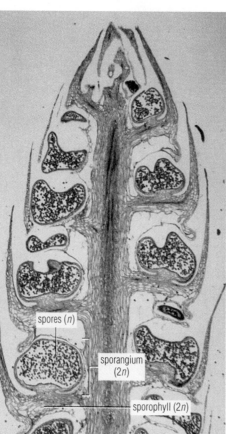

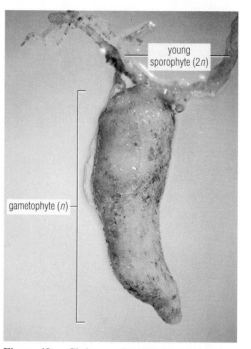

Figure 48a Clubmoss, *Huperzia* (*Lycopodium*) *lucidulum*, **sporophyte.** Sporangia in this species are not organized into distinct cones (= strobili) (live, 1.6×). (Photo by J. W. Perry)

Figure 48d Clubmoss (*Lycopodium*) **cone (strobilus)** (prep. slide, l.s., 20×). (Photo by J. W. Perry)

Figure 48e Clubmoss (*Lycopodium*) **gametophyte.** This gametophyte grows beneath the soil surface (live, 3.2×). (Photo courtesy Dean P. Whittier)

Figure 49a   Spikemoss (*Selaginella paullescens*) **sporophyte** (live, 1/2×). (Photo by J. W. Perry)

Figure 49b   Spikemoss (*Selaginella*) branch with **cones** (strobili). The sporangia of all species of *Selaginella* are organized into cones (live, 1×). (Photo by J. W. Perry)

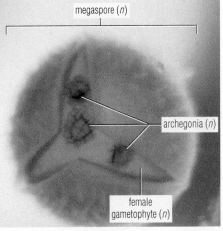

Figure 49d   Mega-spore containing a female gametophyte that has produced archegonia (live, w.m., 70×). (Photo courtesy W. Carl Taylor)

Figure 49c   Spike-moss (*Selaginella*) **cone.** Note that spikemosses are heterosporus, i.e., distinct male spores (**microspores**) and female spores (**megaspores**) are produced (prep. slide, l.s., 20×). (Photo by J. W. Perry)

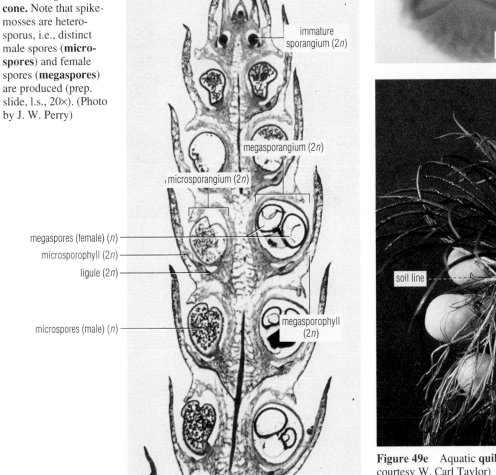

Figure 49e   Aquatic **quillwort** (*Isoetes*) **sporophyte** (live, 1×). (Photo courtesy W. Carl Taylor)

Figure 50b Horsetail (*Equisetum sylvaticum*) sporophyte. Like many other horsetail species, this one produces cones on shoots that are also photosynthetic (live, 1.7×). (Photo by J. W. Perry)

Figure 50c Horsetail (*Equisetum*) gametophyte. No archegonia are present on this gametophyte (prep. slide, w.m., 30×). (Photo by J. W. Perry)

Figure 50a Horsetail (*Equisetum arvense*) sporophytes. This species produces two types of shoots from the same rhizome. The reproductive shoot on the left comes up early in the spring, has very little chlorophyll, and produces a spore-containing **cone (strobilus)** at its tip. The sterile shoot on the right comes up several weeks later, carries out most of the plant's photosynthesis, and never produces cones (live, 1.7×). (Photo by J. W. Perry)

Figure 51a Morphology of a **typical fern** (*Polypodium virginianum*). Middle frond has its lower surface up (herbarium specimen, 1×). (Photo by J. W. Perry)

Figure 51b Fronds of the ostrich fern (*Matteuccia struthiopteris*). Note the tips of the fronds, which are not yet fully elongated, and compare with Figure 51c, a still earlier stage of development of the frond. This fern produces its sporangia on specialized fertile fronds (live, 2/5×). (Photo by J. W. Perry)

Figure 51c Very young **fern frond** just emerging from the soil in early spring. The coiled fronds are called **croziers** or **fiddleheads** (live, 1×). (Photo by J. W. Perry)

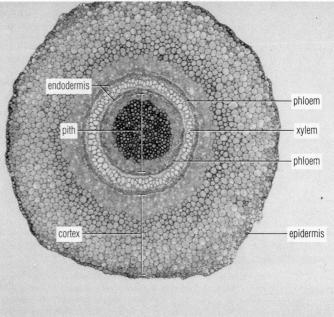

Figure 51d **Fern rhizome** (*Dennstaedtia*) (prep. slide, c.s., 30×). (Photo by J. W. Perry)

Figure 52a Undersurface of **holly fern (*Cyrtomium falcatum*) frond**. Each "dot" is a **sorus** (plural, *sori*) that contains numerous sporangia (live, 3/5×). (Photo by J. W. Perry)

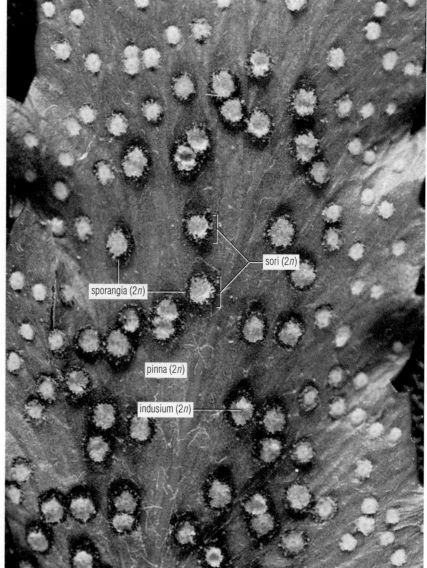

Figure 52b Detail of **holly fern pinna** (leaflet). Brown **sporangia** protrude from the edge of the **indusium**, an umbrella-shaped covering over each sorus (live, 3.8×). (Photo by J. W. Perry)

Figure 52c A single **sorus** of **hare's foot fern (*Polypodium aureum*)**. The sorus of this fern lacks an indusium. Each sporangium has a row of cells with thickened cell walls called the **annulus** (live, w.m., 30×). (Photo by J. W. Perry)

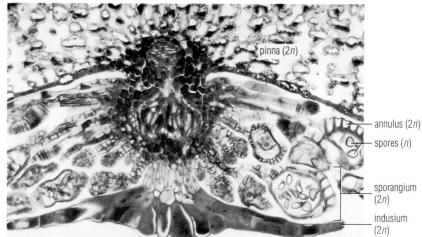

Figure 52d Section through the **sorus** of **holly fern (*Cyrtomium*)** (prep. slide, c.s., 60×). (Photo by J. W. Perry)

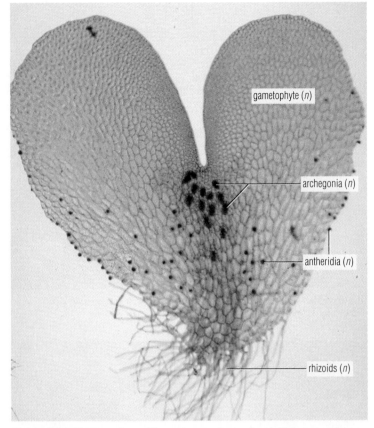

Figure 53a Undersurface of **fern gametophyte**, showing **gametangia (antheridia and archegonia)**. The gametophyte is sometimes called a **prothallus (= prothallium)** (prep. slide, w.m., 20×). (Photo by J. W. Perry)

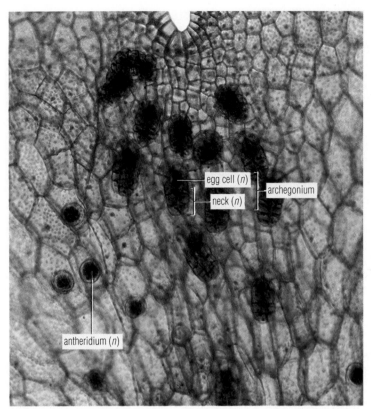

Figure 53b **Gametangia** (antheridia and archegonia) on undersurface of **fern gametophyte** (prep. slide, w.m., 100×). (Photo by J. W. Perry)

Figure 53c Longitudinal section of a **fern gametophyte with antheridia** (prep. slide, l.s., 120×). (Photo by J. W. Perry)

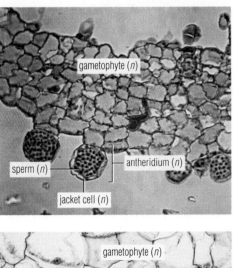

Figure 53d Section of **fern gametophyte with archegonia** (prep. slide, l.s., 250×). (Photo by J. W. Perry)

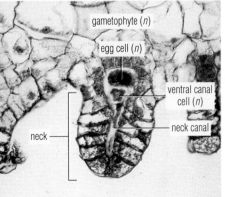

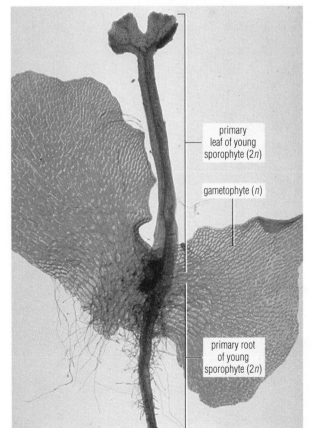

Figure 53e **Fern gametophyte** with an attached **sporophyte** produced by fertilization of an egg in one of the archegonia (prep. slide, w.m., 20×). (Photo by J. W. Perry)

Figure 54a   Sago palm (*Cycas revoluta*) sporophyte (live, 1/25×). (Photo by J. W. Perry)

Figure 54b   Female *Zamia* sporophyte with seed cones (live, 2/5×). (Photo by J. W. Perry)

Figure 54c   Maidenhair tree (*Ginkgo biloba*), the only living species of the phylum. **Sporophyte** (live, 1/100×). (Photo by J. W. Perry)

Figure 54d   A portion of the shoot of *Ginkgo*, showing the very distinctive leaves (live, 1×). (Photo by J. W. Perry)

Figure 55a *Pinus* **(pine) tree**, the **sporophyte** generation (live, 1/25×). (Photo by J. W. Perry)

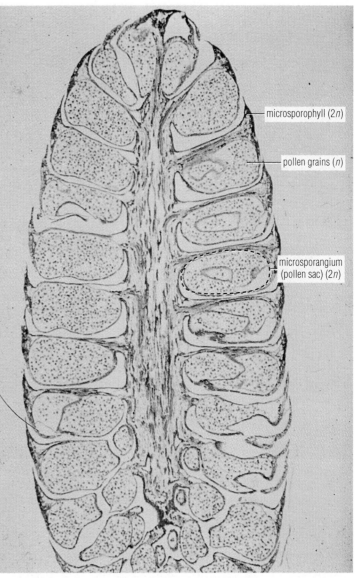

microsporophyll (2*n*)

pollen grains (*n*)

microsporangium (pollen sac) (2*n*)

Figure 55c *Pinus*, **male strobilus** (cone) containing **pollen grains** (immature male gametophytes) (prep. slide, l.s., 20×). (Photo by J. W. Perry)

leaf (needle) (2*n*)

pollen (*n*)

male cones

Figure 55b *Pinus* **(pine)**, cluster of **male cones** (strobili) shedding **pollen** (live, 1.8×). (Photo by J. W. Perry)

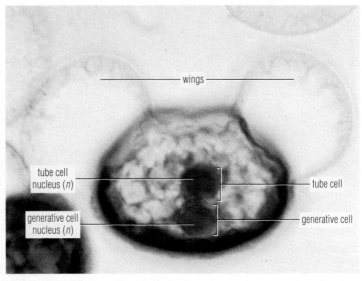

wings

tube cell nucleus (*n*)

tube cell

generative cell nucleus (*n*)

generative cell

Figure 55d *Pinus*, **pollen grain**, the immature male gametophyte (section, 1400×). (Photo by J. W. Perry)

Figure 56a *Pinus,* **female strobili.** At the tip are tiny female cones about two months old; below them is a cluster of cones in their second year of development (about fourteen months old). This species (red pine) requires three seasons for maturation of the cones before the seeds are shed (live, 1/2×). (Photo by J. W. Perry)

Figure 56b **Douglas fir (***Pseudotsuga***),** mature **female strobili.** By contrast with those of pine, these cones have very long sterile bracts (live, 1/2×). (Photo by J. W. Perry)

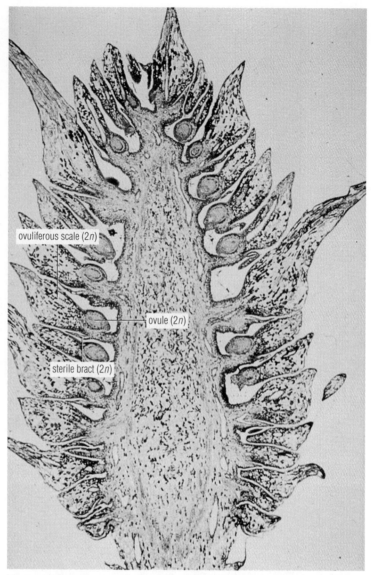

Figure 56c *Pinus,* young (two-month-old) **female strobilus** (prep. slide, l.s., 20×). (Photo by J. W. Perry)

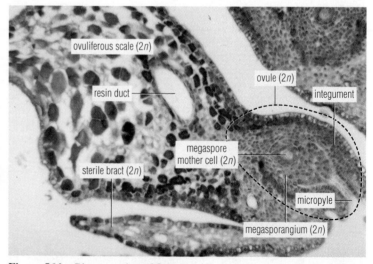

Figure 56d *Pinus,* portion of **female cone** showing **megaspore mother cell** within **megasporangium** (prep. slide, l.s., 100×). (Photo by J. W. Perry)

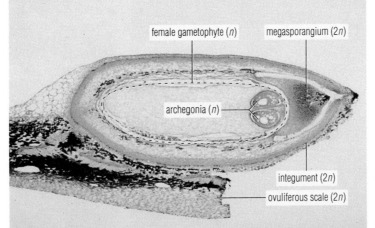

Figure 57a *Pinus*, **ovule** in stage where the **female gametophyte** contains **archegonia** (prep. slide, l.s., 15×). (Photo by J. W. Perry)

female gametophyte (*n*) | megasporangium (2*n*)

archegonia (*n*)

integument (2*n*)

ovuliferous scale (2*n*)

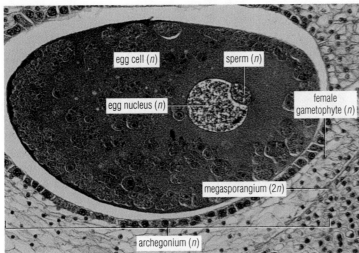

Figure 57b *Pinus*, **archegonium** containing large **egg cell** at time of **fertilization** (prep. slide, l.s., 100×). (Photo by J. W. Perry)

egg cell (*n*) | sperm (*n*)

egg nucleus (*n*)

female gametophyte (*n*)

megasporangium (2*n*)

archegonium (*n*)

winged seed (*n* + 2*n*)

wing of seed

ovuliferous scale (2*n*)

Figure 57c *Pinus*, mature **female cones** and a single **seed** (live, 1/3×). (Photo by J. W. Perry)

seed coat (2*n*)

cotyledons (2*n*)

new photosynthetic shoot of sporophyte (2*n*)

hypocotyl (2*n*)

Figure 57e *Pinus* (**pinyon pine**) **seedling** with **seed coat** still attached to **cotyledons** (live, 1×). (Photo by J. W. Perry)

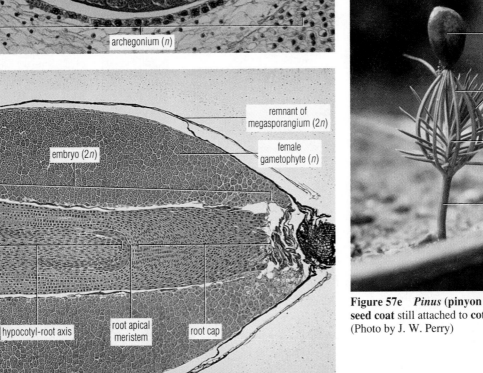

remnant of megasporangium (2*n*)

embryo (2*n*)

female gametophyte (*n*)

cotyledons | shoot apical meristem (epicotyl) | hypocotyl-root axis | root apical meristem | root cap

Figure 57d *Pinus*, **seed** containing **embryo**. The hard seed coat was removed during slide preparation (prep. slide, l.s., 30×). (Photo courtesy Biodisc, Inc.)

Figure 58a   Ephedra. This plant is commonly called Mormon tea and grows in the southwestern United States (live, 1/20×). (Photo by J. W. Perry)

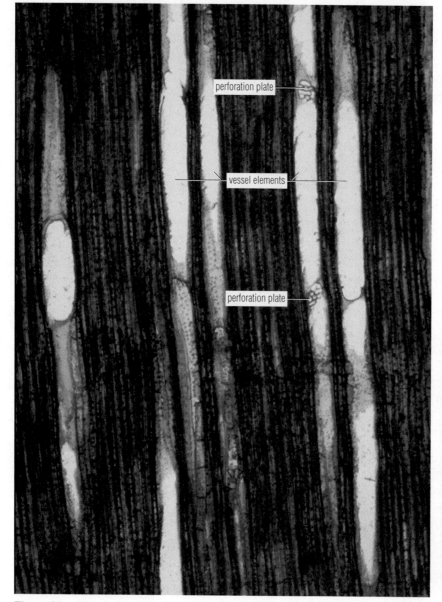

Figure 58b   *Gnetum leyboldii*, a species that grows as a vine. All *Gnetum* species have angiosperm-like leaves (live, 1/30×). (Photo by J. W. Perry)

Figure 58c   Wood of a gnetophyte (*Ephedra*) showing **xylem vessels**, the characteristic that gives the division the common name "vessel-containing gymnosperms" (prep. slide, l.s., 650×). (Photo by J. W. Perry)

Figure 58d   *Gnetum gnemon*, a species with a treelike habit (live, 1/30×). (Photo by J. W. Perry)

Figure 59a **Monocot flower** (*Trillium*). Typical of monocotyledons, the floral parts are in threes (live, 1×). (Photo by J. W. Perry)

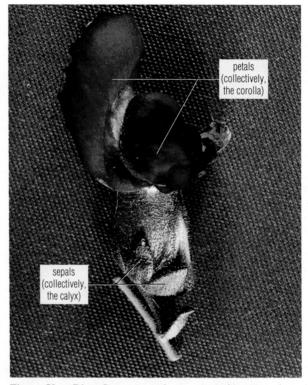

Figure 59c **Dicot flower** (snapdragon, *Antirrhinum*), an irregular (bilaterally symmetrical) flower (live, 1.5×). (Photo by J. W. Perry)

Figure 59b **Dicot flowers** (spring beauty, *Claytonia*), a regular (radially symmetrical) flower. Dicotyledon flowers typically have their parts in fours or fives (live, 1.4×). (Photo by J. W. Perry)

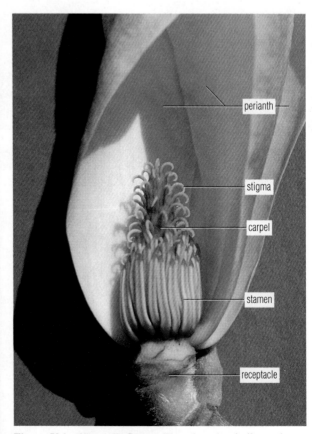

Figure 59d *Magnolia* **flower.** Thought to be similar to the flowers that evolved in the Cretaceous period (between 144 and 65 million years ago) this dicot flower lacks distinction between sepals and petals, has its floral parts arranged in a spiral pattern on the stem, and has numerous stamens and carpels (pistils) (live, 1×). (Photo by J. W. Perry)

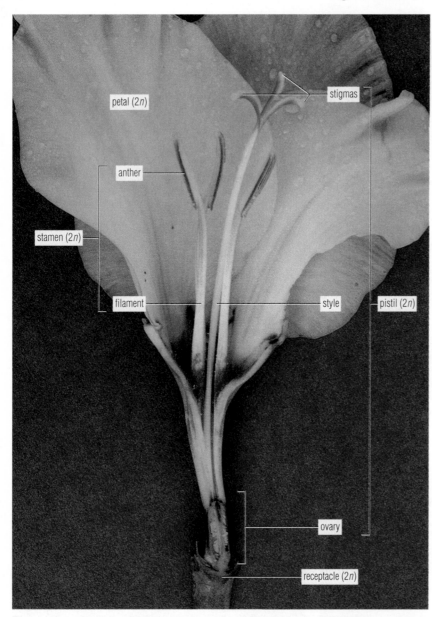

Figure 60a *Gladiolus* **flower**, a monocot (live, 1.3×). (Photo by J. W. Perry)

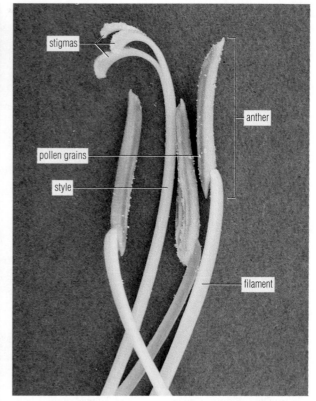

Figure 60b Upper portion of **stamens and pistil** of *Gladiolus* (live, 1.8×). (Photo by J. W. Perry)

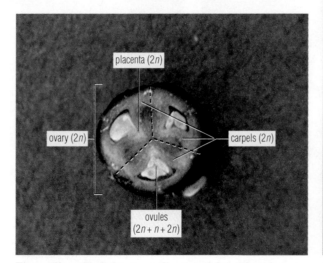

Figure 60c *Gladiolus* **ovary** (live, c.s., 3.2×). (Photo by J. W. Perry)

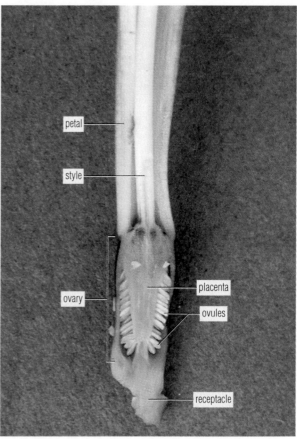

Figure 60d Lower portion of *Gladiolus* **pistil** (live, l.s., 1.9×). (Photo by J. W. Perry)

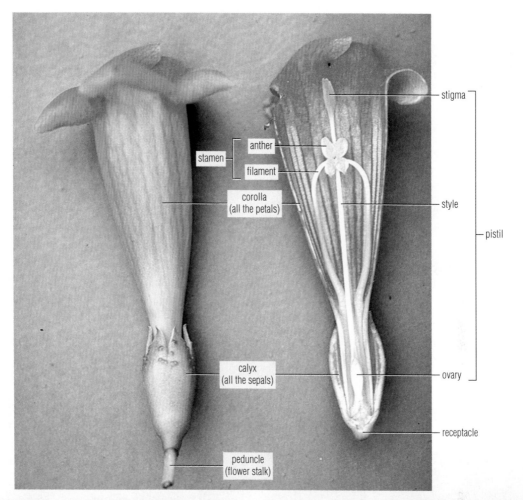

Figure 61a   Trumpet creeper (*Campsis radicans*) flowers. This **dicot flower** is **hypogynous**, with all floral parts clearly originating below the ovary, which is superior. Both the five sepals and the five petals are fused in this regular, complete flower (live, 1.3×). (Photo by J. W. Perry)

stamen
anther
filament
corolla (all the petals)
stigma
style
pistil
ovary
calyx (all the sepals)
receptacle
peduncle (flower stalk)

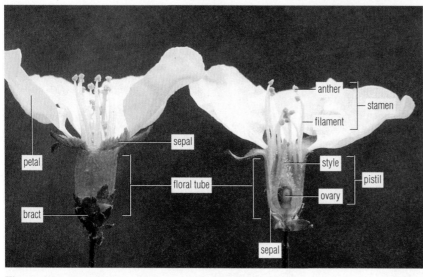

petal
sepal
anther
stamen
filament
style
pistil
floral tube
ovary
bract
sepal

Figure 61b   Cherry (*Prunus*) flowers. This **dicot flower** is **perigynous**, with its floral parts fused together, forming a floral tube. Hence, the petals and stamens appear to arise at the top of this floral tube. The ovary is superior. The flower is regular and complete. Its ovary has five separate styles, an indication that the pistil consists of five carpels (live, 1.8×). (Photo by J. W. Perry)

stigmas
styles
pistil
ovule
ovary
peduncle

Figure 61c   Apple (*Pyrus malus*) flower. This **dicot flower** is **epigynous**, with the floral parts coming off above the ovary, which is inferior. The flower is regular and complete (live, 1.8×). (Photo by J. W. Perry)

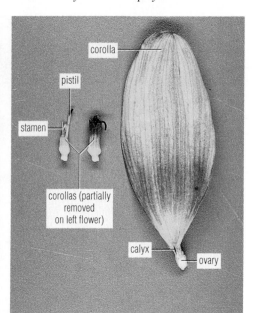

Figure 62b   Two **disk flowers** and a **ray flower** from a sunflower inflorescence (cluster of flowers) (live, 4/5×). (Photo by J. W. Perry)

Figure 62a   **Composite inflorescence** (cluster of flowers). This sunflower (*Helianthus*) consists of ray flowers and disk flowers (live, 2/5×). (Photo by J. W. Perry)

Figure 62c   **Inflorescence** (cluster of flowers) of timothy **grass** (*Phelum*). Anthers are abundant at this stage of development (live, 1×). (Photo by J. W. Perry)

Figure 62d   **Lily (*Lilium*) flower**. Sections of floral buds (unexpanded flowers) are used to create the slides depicted in Figures 63a through 63c and 64a through 64e (live, 1.3×). (Photo by J. W. Perry)

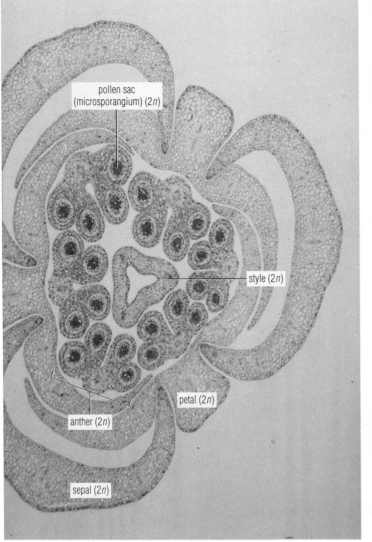

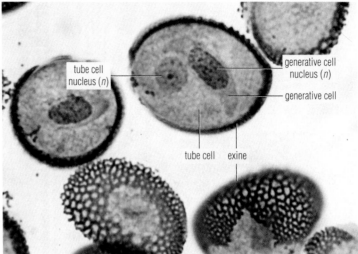

Figure 63c   **Lily pollen** (prep. slide, c.s., 300×). (Photo by J. W. Perry)

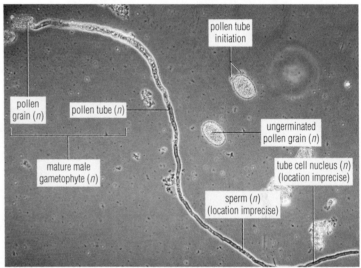

Figure 63a   **Lily bud.** Each anther consists of four fused **microsporangia** (pollen sacs) and at this stage of development contains diploid microspore mother cells (prep. slide, c.s., 40×). (Photo by J. W. Perry)

Figure 63d   Germinating **pollen grain** of *Impatiens* (live, w.m., 150×). (Photo by J. W. Perry)

Figure 63b   **Lily anther** containing **pollen grains** (prep. slide, c.s., 40×). (Photo courtesy Biodisc, Inc.)

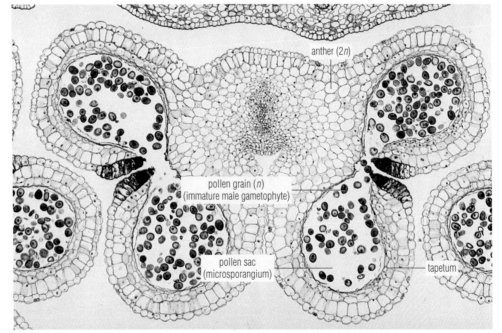

Figure 64b **Lily ovule** containing **megaspore mother cell** within the megasporangium. The ovule in this figure is less developed than the ovules in Figure 64a (prep. slide, c.s., 150×). (Photo by J. W. Perry)

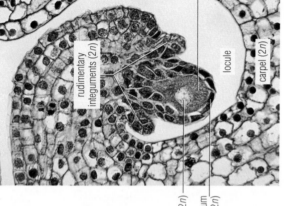

megaspore mother cell (2n)

rudimentary integuments (2n)

locule

carpel (2n)

nucleus (2n)

megasporangium (nucellus) (2n)

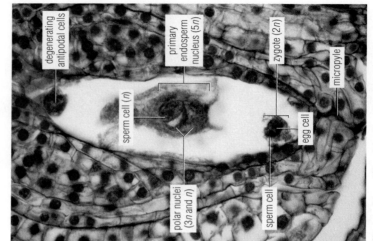

degenerating antipodal cells

primary endosperm nucleus (5n)

zygote (2n)

micropyle

sperm cell (n)

egg cell

polar nuclei (3n and n)

sperm cell

Figure 64e **"Double fertilization"** in lily gametophyte. In actuality, there is only one true fertilization by fusion of one sperm and the egg cell. The other sperm fuses with the two central cell (polar) nuclei (prep. slide, c.s., 250×). (Photo by J. W. Perry)

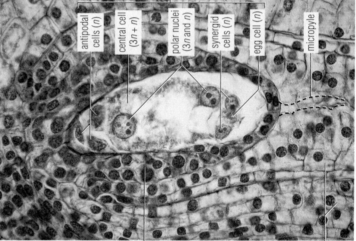

female gametophyte (embryo sac)

antipodal cells (n)

central cell (3n + n)

polar nuclei (3n and n)

synergid cells (n)

egg cell (n)

micropyle

megasporangium (nucellus) (2n)

integuments (2n)

Figure 64d **Mature (seven-celled, eight-nucleate) female gametophyte** of lily (prep. slide, c.s., 250×). (Photo by J. W. Perry)

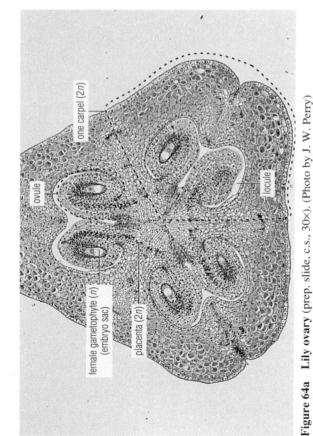

one carpel (2n)

ovule

locule

female gametophyte (n) (embryo sac)

placenta (2n)

Figure 64a **Lily ovary** (prep. slide, c.s., 30×). (Photo by J. W. Perry)

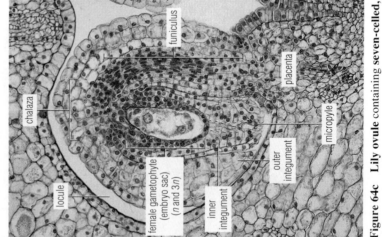

funiculus

chalaza

placenta

locule

female gametophyte (embryo sac) (n and 3n)

inner integument

outer integument

micropyle

Figure 64c **Lily ovule** containing **seven-celled, eight-nucleate female gametophyte.** The gametophyte is sometimes referred to as an embryo sac (prep. slide, c.s., 90×). (Photo by J. W. Perry)

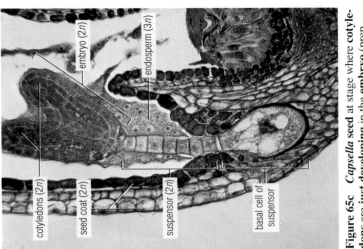

embryo (2n)

endosperm (3n)

cotyledons (2n)

seed coat (2n)

suspensor (2n)

basal cell of suspensor

Figure 65c *Capsella* seed at stage where **cotyledons are just developing** in the **embryo** (prep. slide, l.s., 200×). (Photo courtesy Biodisc, Inc.)

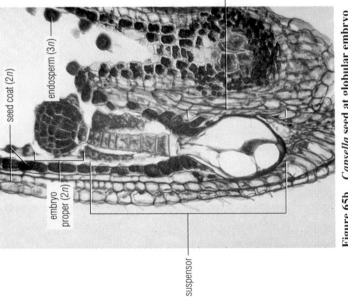

basal cell of suspensor

endosperm (3n)

seed coat (2n)

embryo proper (2n)

suspensor

Figure 65b *Capsella* seed at globular embryo proper stage (prep. slide, l.s., 200×). (Photo courtesy Biodisc, Inc.)

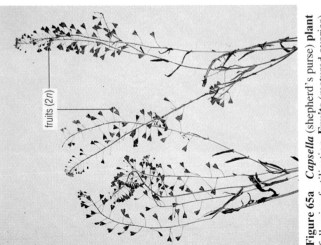

fruits (2n)

Figure 65a *Capsella* (shepherd's purse) **plant** following fertilization. **Fruits** (matured ovaries) have developed (herbarium specimen, 1/3×). (Photo by J. W. Perry)

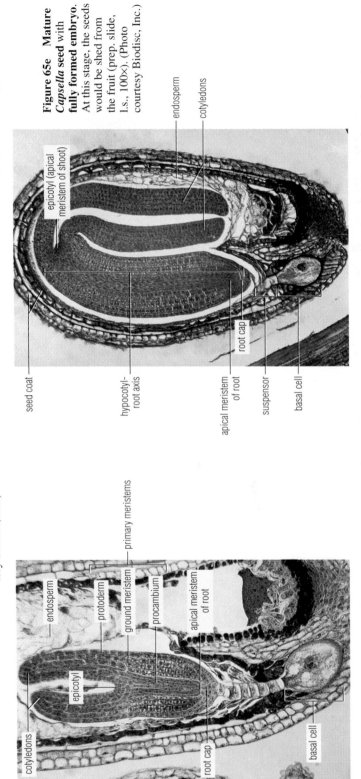

Figure 65e Mature *Capsella* **seed** with **fully formed embryo.** At this stage, the seeds would be shed from the fruit (prep. slide, l.s., 100×). (Photo courtesy Biodisc, Inc.)

endosperm

cotyledons

epicotyl (apical meristem of shoot)

seed coat

hypocotyl–root axis

apical meristem of root

root cap

suspensor

basal cell

primary meristems

endosperm

protoderm

ground meristem

procambium

apical meristem of root

cotyledons

epicotyl

root cap

basal cell

Figure 65d *Capsella* **seed** at stage where **cotyledons** are more fully developed in the **embryo** (prep. slide, l.s., 150×). (Photo courtesy Biodisc, Inc.)

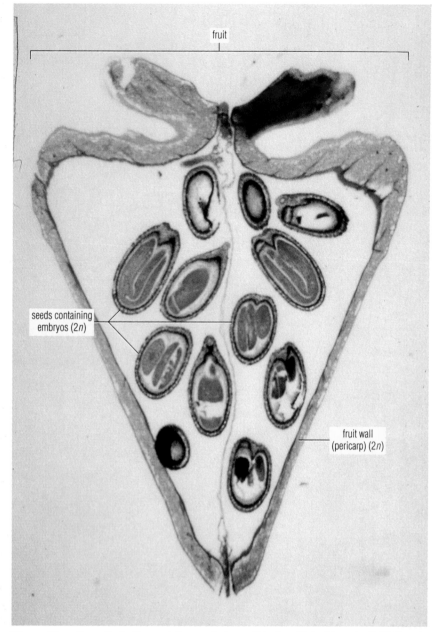

Figure 66a   *Capsella* **fruit** containing **seeds** (prep slide, l.s., 80×). (Photo by J. W. Perry)

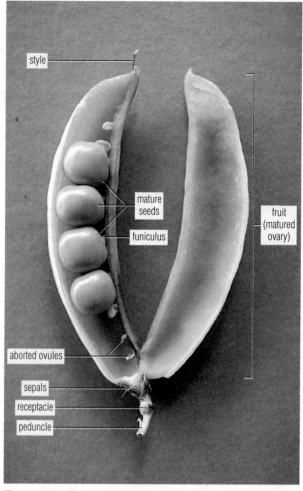

Figure 66b   **Fruit** (a **pod**) containing seeds (matured ovules) of **garden pea** (live, 1×). (Photo by J. W. Perry)

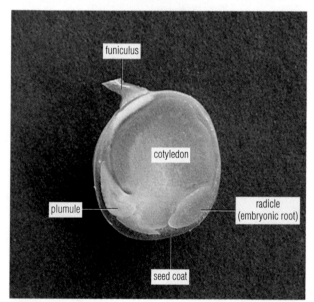

Figure 66c   **Garden pea seed** (live, sec., 3×). (Photo by J. W. Perry)

Figure 67a   Germinating bean seed, about one day after sowing (live, 2.2×). (Photo by J. W. Perry)

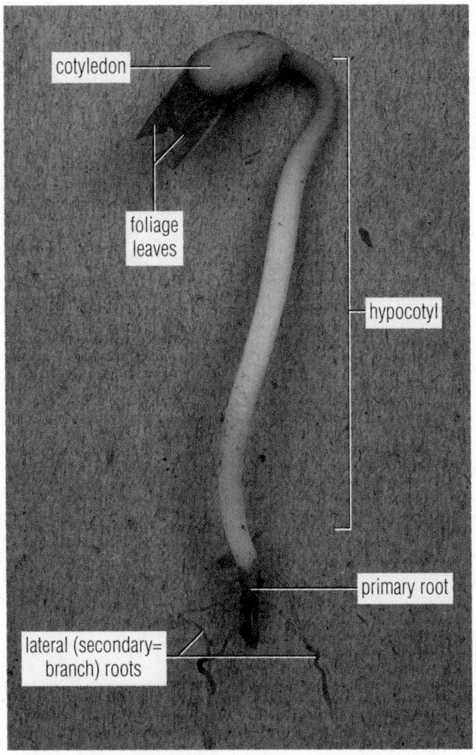

Figure 67b   Germinating bean seed, three or four days after sowing. Seed coat has been shed (live, 1×). (Photo by J. W. Perry)

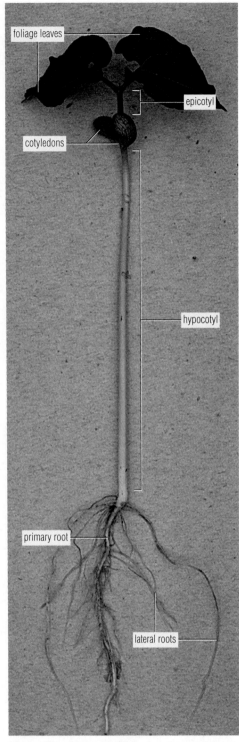

Figure 67c   Bean seedling, about seven days after sowing seed (live, 3/5×). (Photo by J. W. Perry)

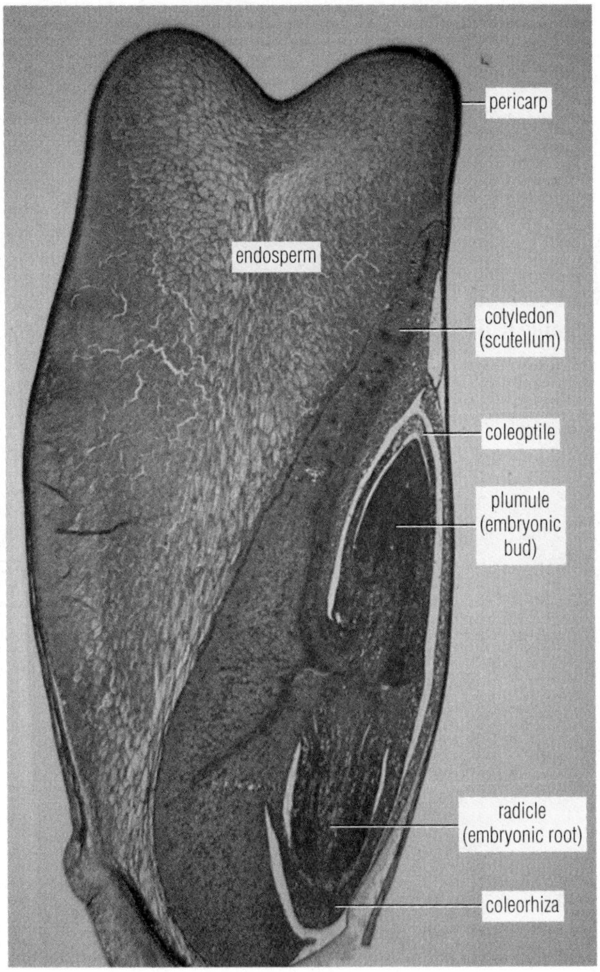

Figure 67d   Corn grain. A grain is the one-seeded fruit of a grass. The seed coat is fused to, and virtually impossible to discriminate from, the fruit wall (pericarp) (prep. slide, l.s., 7.5×). (Photo by J. W. Perry)

Figure 68a **Immature tomato fruit,** a **berry** (live, 1×). (Photo by J. W. Perry)

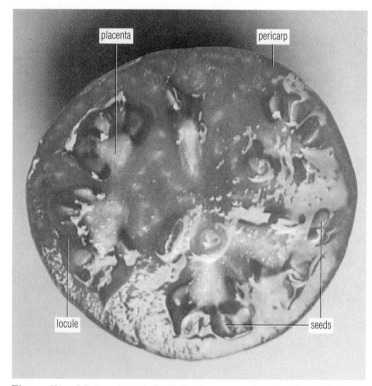

Figure 68b **Mature tomato fruit.** During maturation, the placental tissue breaks down, leaving a watery mass (live, c.s, 1×). (Photo by J. W. Perry)

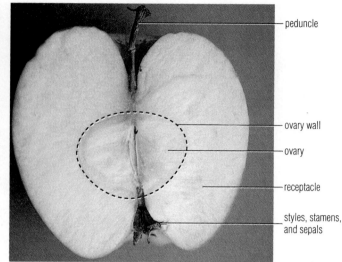

Figure 68c **Apple fruit,** a **pome.** The receptacle of the flower contributes to the mass of the fruit, making this an "accessory fruit" (live, l.s., 4/5×). (Photo by J. W. Perry)

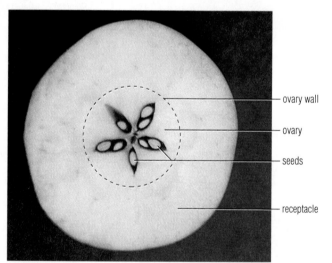

Figure 68d **Apple fruit** (live, c.s., 4/5×). (Photo by J. W. Perry)

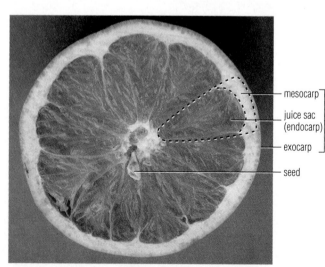

Figure 68e **Grapefruit fruit,** a **hesperidium.** The structure of an orange is identical (live, c.s., 1/2×). (Photo by J. W. Perry)

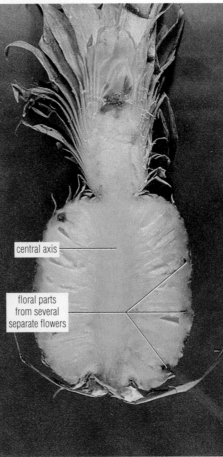

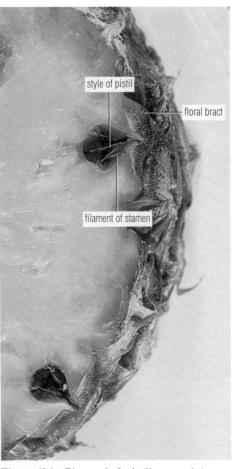

Figure 69a   Black-berry fruits, aggregate fruits consisting of numerous small subunits called fruitlets (live, 1×). (Photo by J. W. Perry)

Figure 69b   Pineapple, a multiple fruit formed from a cluster of closely grouped flowers (live, 2/5×). (Photo by J. W. Perry)

Figure 69c   Pineapple fruit (live, l.s., 2/5×). (Photo by J. W. Perry)

Figure 69d   Pineapple fruit (live, c.s., 1×). (Photo by J. W. Perry)

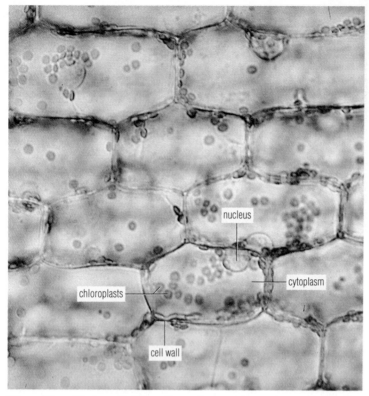

Figure 70a *Elodea* **cells**. The cell membrane and vacuolar membrane are too thin to be resolved with the light microscope. These cells are **turgid** (live, w.m., 600×). (Photo by J. W. Perry)

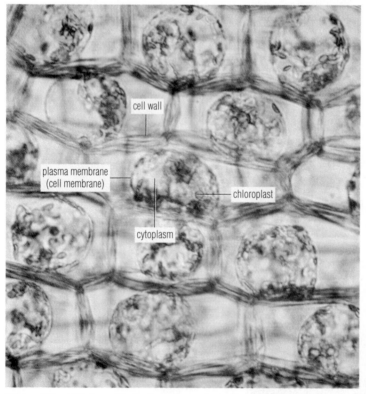

Figure 70b *Elodea* **cells** that have been **plasmolyzed** by placing them in a hypertonic solution (live, w.m., 600×). (Photo by J. W. Perry)

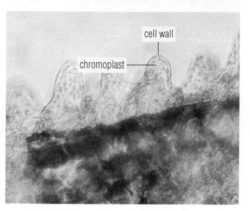

Figure 70c Red-onion bulb (leaf) **parenchyma cells**. In some cells the vacuole is filled with a red **anthocyanin pigment**; in others it has ruptured, releasing the pigment (live, w.m., 250×). (Photo by J. W. Perry)

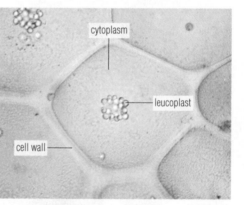

Figure 70d Flower-petal **parenchyma cells** of *Primula kewensis* containing yellow **chromoplasts** (live, w.m., 400×). (Photo by J. W. Perry)

Figure 70e Leaf cells of *Zebrina* with **leucoplasts** clustered around the nucleus. The faint coloration is due to anthocyanin pigment in each cell's vacuole (live, w.m., 400×). (Photo by J. W. Perry)

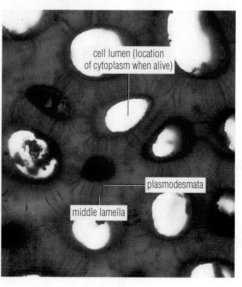

Figure 70f **Plasmodesmata** in thick walls of parenchyma cells from the endosperm of *Diospyros* (persimmon). The **middle lamella** cementing together adjacent cells is perpendicular to the fine plasmodesmata (prep. slide, c.s., 400×). (Photo by J. W. Perry)

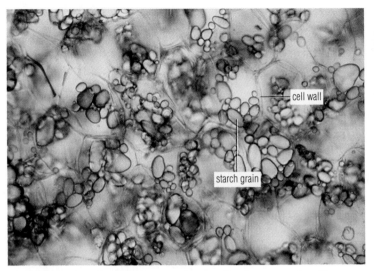

Figure 71a   **Starch grains** in parenchyma cells of a potato tuber (live, sec., 150×). (Photo by J. W. Perry)

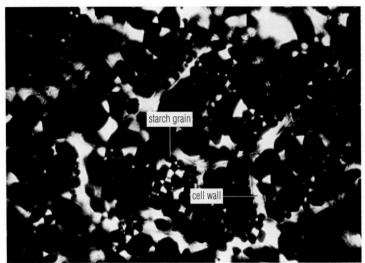

Figure 71b   **Starch grains** as stained with iodine (I₂KI) solution (live, sec., 150×). (Photo by J. W. Perry)

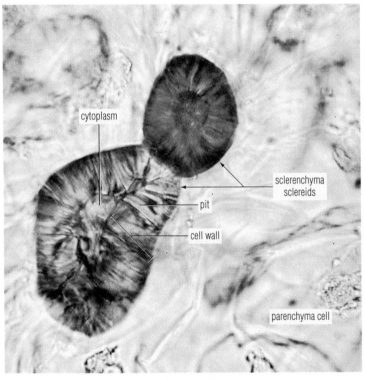

Figure 71c   **Sclerenchyma sclereid cells** from the flesh of pear fruit, stained with phloroglucinol in 20% hydrochloric acid. Lines in cell walls are pits (channels) in the secondary cell wall (live, w.m., 600×). (Photo by J. W. Perry)

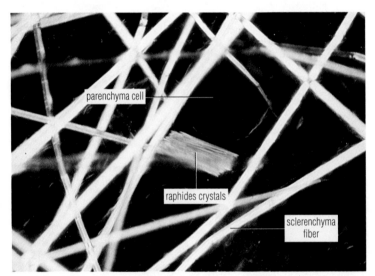

Figure 71d   **Sclerenchyma fibers** and raphides crystals, as seen with polarized light (live, w.m., 600×). (Photo by J. W. Perry)

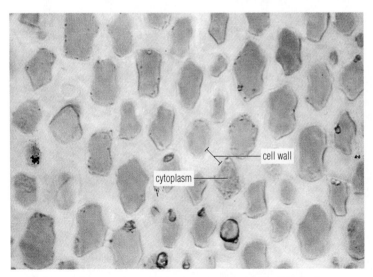

Figure 71e   **Collenchyma cells** from the petiole (leaf stalk) of celery. Darkened areas are the cell lumens (regions where cytoplasm is located), white areas the cell walls (live, c.s., 600×). (Photo by J. W. Perry)

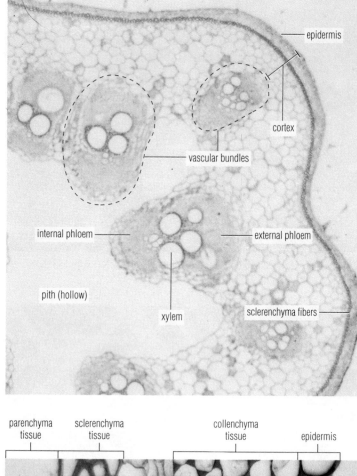

Figure 72a　Prepared slide of ***Cucurbita*** **(squash) stem** (c.s., 70×). (Photo by J. W. Perry)

epidermis

cortex

vascular bundles

internal phloem

external phloem

pith (hollow)

xylem

sclerenchyma fibers

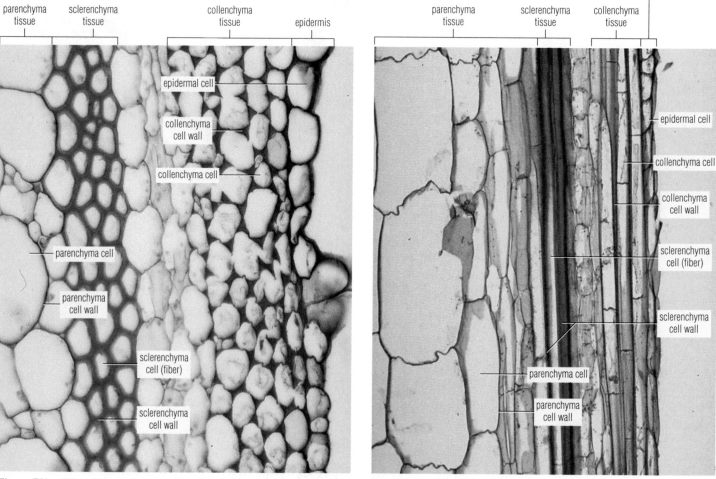

parenchyma tissue

sclerenchyma tissue

collenchyma tissue

epidermis

epidermal cell

collenchyma cell wall

collenchyma cell

parenchyma cell

parenchyma cell wall

sclerenchyma cell (fiber)

sclerenchyma cell wall

parenchyma tissue

sclerenchyma tissue

collenchyma tissue

epidermis

epidermal cell

collenchyma cell

collenchyma cell wall

sclerenchyma cell (fiber)

sclerenchyma cell wall

parenchyma cell

parenchyma cell wall

Figure 72b　Edge of *Cucurbita* stem showing all **three plant cell types** (prep. slide, c.s., 350×). (Photo by J. W. Perry)

Figure 72c　Longitudinal section of edge of *Cucurbita* stem showing **three plant cell types** (prep. slide, l.s., 140×). (Photo by J. W. Perry)

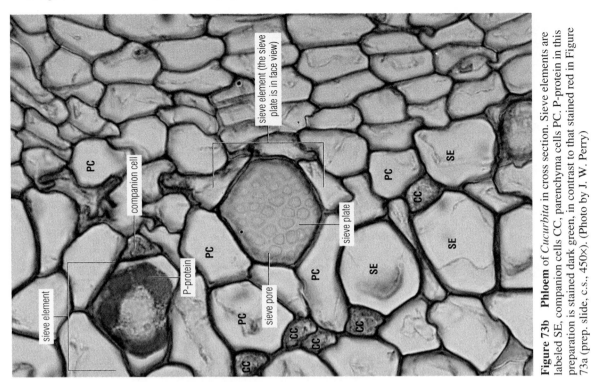

sieve element (the sieve plate is in face view)

companion cell

PC

sieve element

P-protein

PC

sieve pore

PC

PC

PC

CC

CC

CC

sieve plate

SE

SE

PC

CC

SE

Figure 73b **Phloem** of *Cucurbita* in cross section. Sieve elements are labeled SE, companion cells CC, parenchyma cells PC. P-protein in this preparation is stained dark green, in contrast to that stained red in Figure 73a (prep. slide, c.s., 450×). (Photo by J. W. Perry)

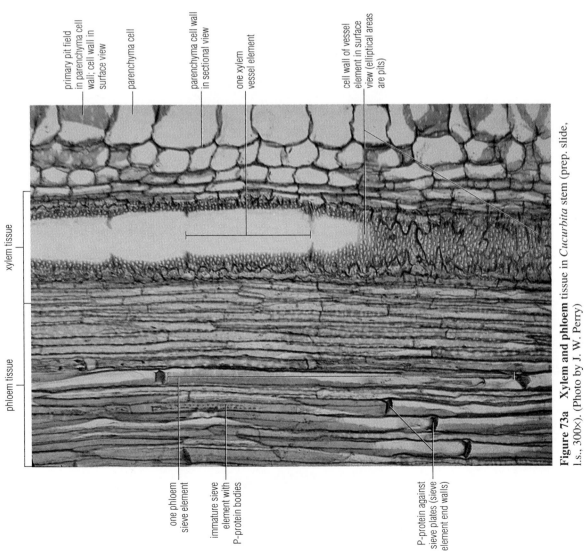

primary pit field in parenchyma cell wall; cell wall in surface view

parenchyma cell

parenchyma cell wall in sectional view

one xylem vessel element

cell wall of vessel element in surface view (elliptical areas are pits)

xylem tissue

phloem tissue

one phloem sieve element

immature sieve element with P-protein bodies

P-protein against sieve plates (sieve element end walls)

Figure 73a **Xylem and phloem** tissue in *Cucurbita* stem (prep. slide, l.s., 300×). (Photo by J. W. Perry)

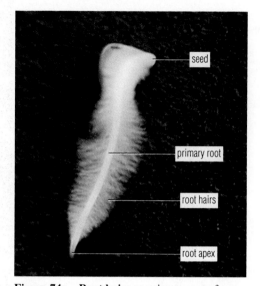

seed

primary root

root hairs

root apex

Figure 74a   Root hairs on primary root of germinating **radish** seed (live, 5×). (Photo by J. W. Perry)

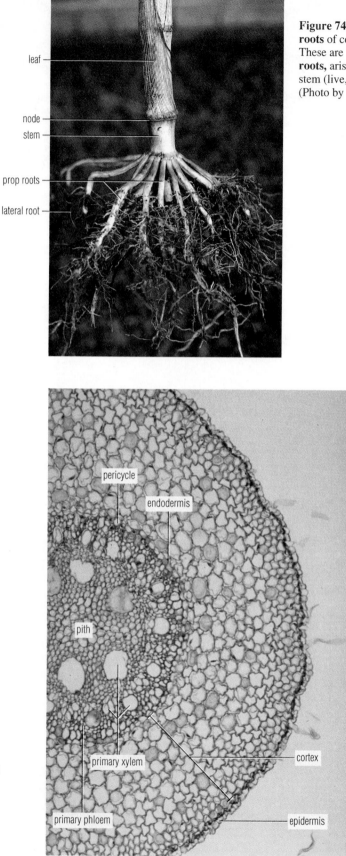

Figure 74c   Prop roots of corn (*Zea*). These are **adventitious roots,** arising from the stem (live, 1/3×). (Photo by J. W. Perry)

leaf

node

stem

prop roots

lateral root

pericycle

endodermis

pith

primary xylem

primary phloem

cortex

epidermis

Figure 74d   Monocot (corn, *Zea*) **root** (prep. slide, c.s., 120×). (Photo by J. W. Perry)

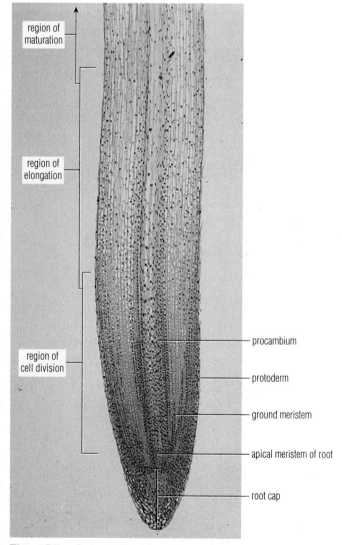

region of maturation

region of elongation

region of cell division

procambium

protoderm

ground meristem

apical meristem of root

root cap

Figure 74b   Onion (*Allium*) **root** (prep. slide, l.s., 30×). (Photo by J. W. Perry)

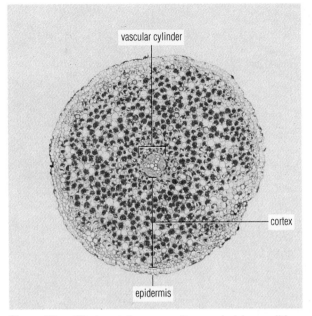

Figure 75a   Dicot root (buttercup, *Ranunculus*) (prep. slide, c.s., 60×). (Photo by J. W. Perry)

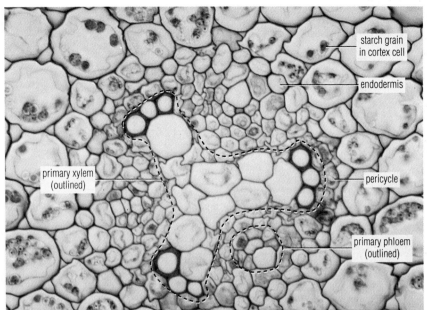

Figure 75b   Vascular cylinder of **immature dicot root** (buttercup, *Ranunculus*) (prep. slide, c.s., 500×). (Photo by J. W. Perry)

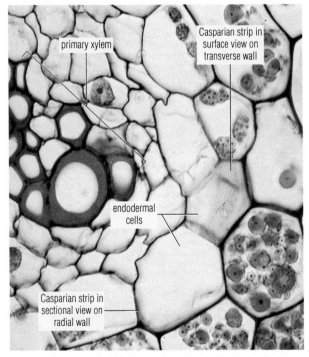

Figure 75d   Casparian strip in **root endodermal cells** (prep. slide, c.s., 500×). (Photo by J. W. Perry)

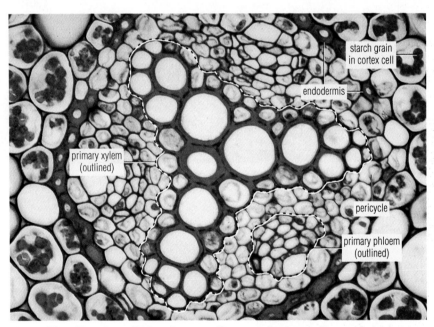

Figure 75c   Vascular cylinder of **mature dicot root** (buttercup, *Ranunculus*) (prep. slide, c.s., 500×). (Photo by J. W. Perry)

Figure 76a **Adventitious roots** arising from stem cutting of geranium plant (live, 1×). (Photo by J. W. Perry)

Figure 76b **Lateral (branch = secondary) roots** arising from primary roots of water lettuce (*Pistia*) (live, 1×). (Photo by J. W. Perry)

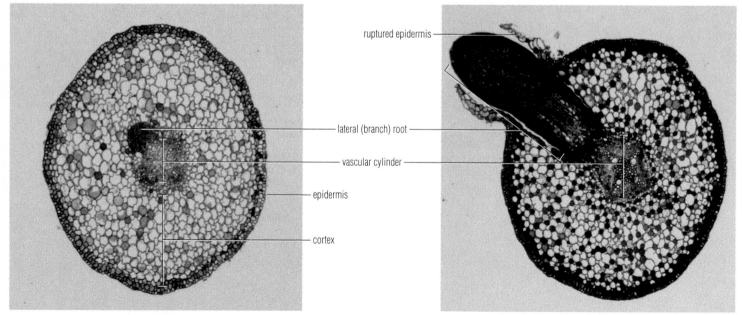

Figure 76c **Branch root** arising from pericycle of a willow (*Salix*) root, initial stage (prep. slide, c.s., 65×). (Photo by J. W. Perry)

Figure 76d **Branch root** that has broken through the primary root's surface (willow) (prep. slide, c.s., 65×). (Photo by J. W. Perry)

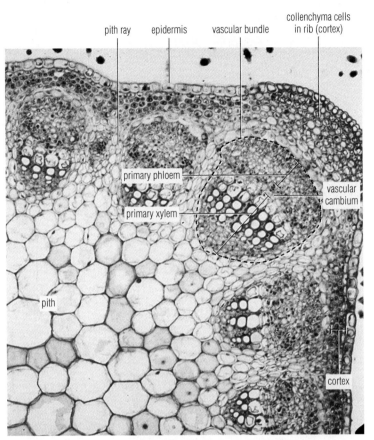

Figure 77a   **Herbaceous dicot** (alfalfa, *Medicago*) **stem** (prep. slide, c.s., 150×). (Photo by J. W. Perry)

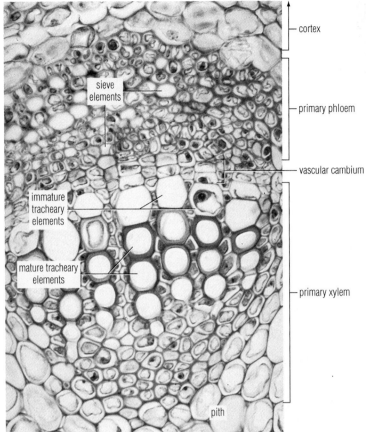

Figure 77b   **Vascular bundle** of **herbaceous dicot** (alfalfa, *Medicago*) **stem** (prep. slide, c.s., 450×). (Photo by J. W. Perry)

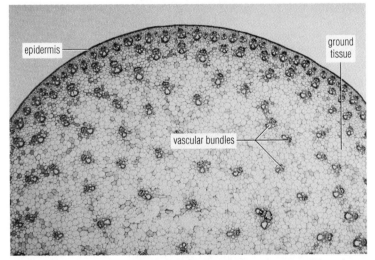

Figure 77c   **Monocot** (corn, *Zea*) **stem** (prep. slide, c.s., 15×). (Photo by J. W. Perry)

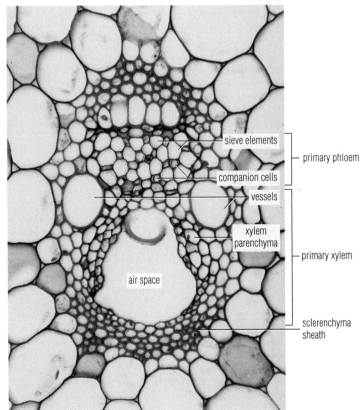

Figure 77d   **Vascular bundle** of **monocot** (corn, *Zea*) **stem** (prep. slide, c.s., 300×). (Photo by J. W. Perry)

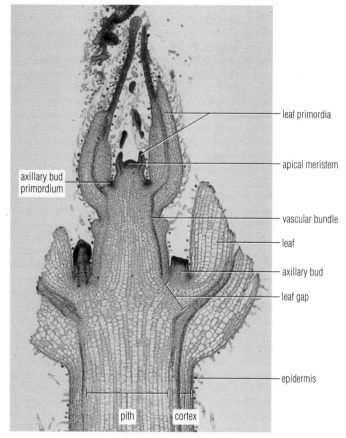

leaf primordia

apical meristem

axillary bud primordium

vascular bundle

leaf

axillary bud

leaf gap

epidermis

pith   cortex

Figure 78a   *Coleus* **shoot tip** (prep. slide, l.s., 30×). (Photo by J. W. Perry)

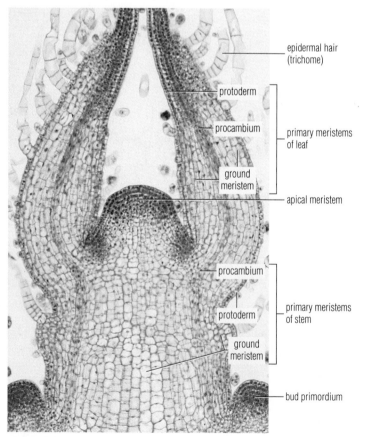

epidermal hair (trichome)

protoderm

procambium

primary meristems of leaf

ground meristem

apical meristem

procambium

primary meristems of stem

protoderm

ground meristem

bud primordium

Figure 78b   *Coleus* **shoot apex** (prep. slide, l.s., 120×). (Photo by J. W. Perry)

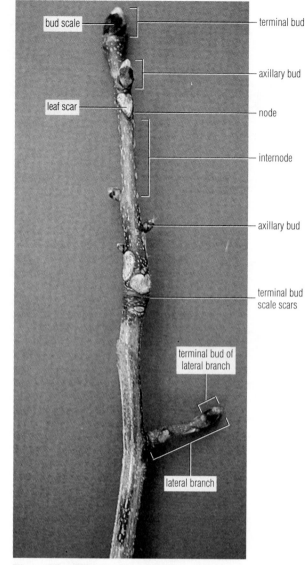

bud scale

terminal bud

axillary bud

leaf scar

node

internode

axillary bud

terminal bud scale scars

terminal bud of lateral branch

lateral branch

Figure 78c   **Woody dicot** (hickory, *Carya*) **branch** (live, 1/2×). (Photo by J. W. Perry)

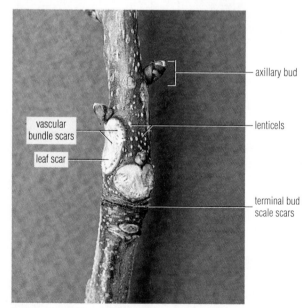

axillary bud

lenticels

vascular bundle scars

leaf scar

terminal bud scale scars

Figure 78d   **Woody dicot** (hickory, *Carya*) **branch** (live, 4/5×). (Photo by J. W. Perry)

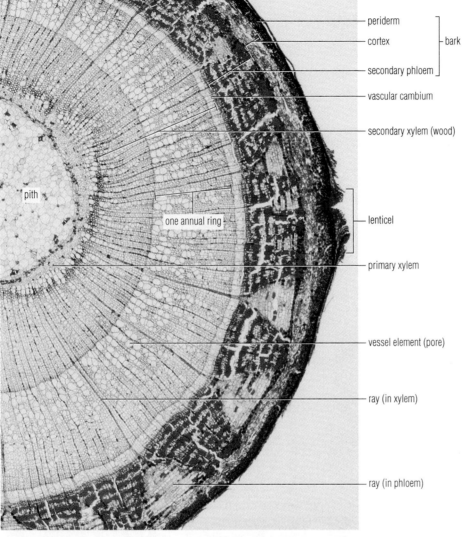

- periderm ⎤
- cortex ⎬ bark
- secondary phloem ⎦
- vascular cambium
- secondary xylem (wood)
- pith
- one annual ring
- lenticel
- primary xylem
- vessel element (pore)
- ray (in xylem)
- ray (in phloem)

Figure 79a   Woody dicot stem (basswood, *Tilia*) that is three years old. The most recently formed annual ring is much narrower than the first two (prep. slide, c.s., 20×). (Photo by J. W. Perry)

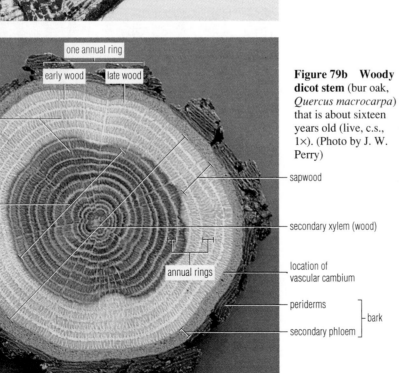

- one annual ring
- early wood
- late wood
- ray
- heartwood
- sapwood
- secondary xylem (wood)
- annual rings
- location of vascular cambium
- periderms ⎤ bark
- secondary phloem ⎦

Figure 79b   Woody dicot stem (bur oak, *Quercus macrocarpa*) that is about sixteen years old (live, c.s., 1×). (Photo by J. W. Perry)

- lenticels

Figure 79c   Cherry (*Prunus*) **bark** with prominent **lenticels** (live, 1/3×). (Photo by J. W. Perry)

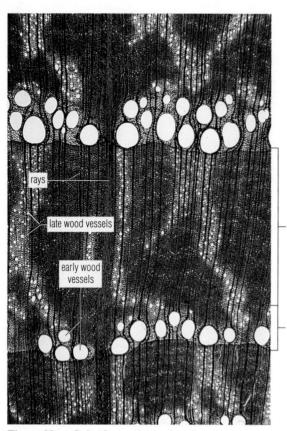

Figure 80a   Oak (*Quercus*) **wood, a ring-porous wood** (prep. slide, c.s., 30×). (Photo by J. W. Perry)

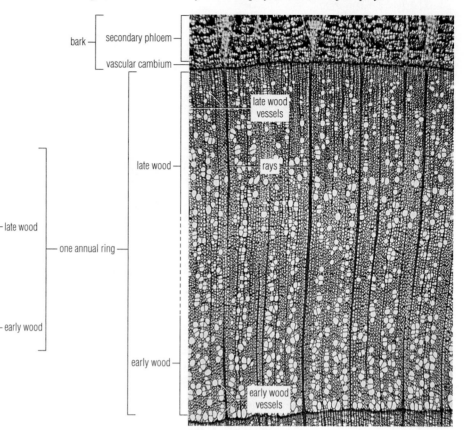

Figure 80b   Basswood (*Tilia*) **wood, a diffuse-porous wood.** The transition from early wood to late wood is much less obvious than in oak (prep. slide, c.s., 30×). (Photo by J. W. Perry)

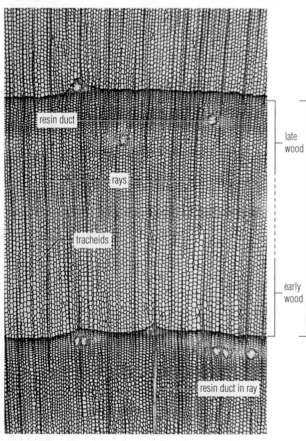

Figure 80c   Pine (*Pinus*) **wood.** Lacking vessels, this is **nonporous wood** (prep. slide, c.s., 30×). (Photo by J. W. Perry)

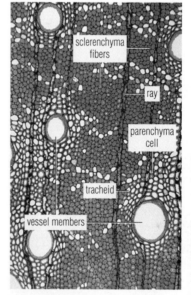

Figure 80d   Oak wood (*Quercus*) (prep. slide, c.s., 60×). (Photo by J. W. Perry)

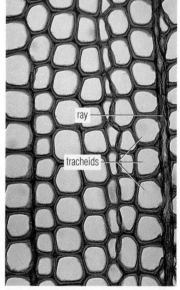

Figure 80e   Pine (*Pinus*) **wood** (prep. slide, c.s., 300×). (Photo by J. W. Perry)

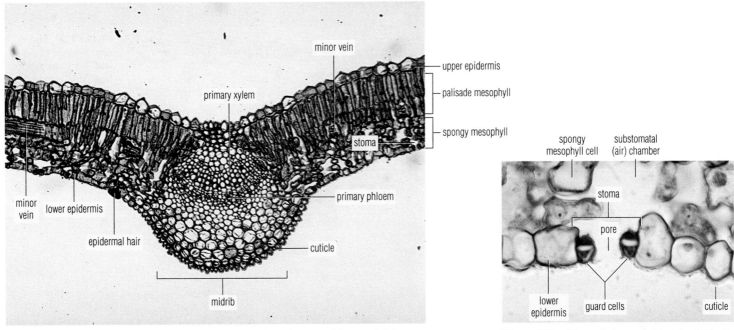

Figure 81a   **Dicot** (lilac, *Syringa*) **leaf, a mesomorphic leaf** (prep. slide, c.s., 30×). (Photo by J. W. Perry)

Figure 81b   **Stoma** in lower epidermis of **dicot** (lilac, *Syringa*) **leaf** (prep. slide, c.s., 600×). (Photo by J. W. Perry)

Figure 81c   Dicot leaf epidermis (prep. slide, w.m., 120×). (Photo by J. W. Perry)

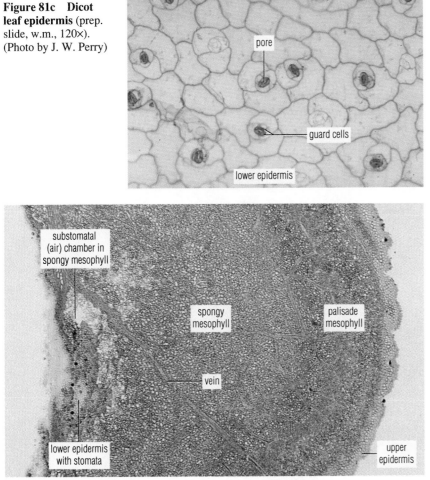

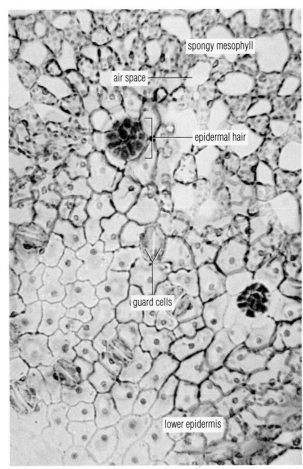

Figure 81d   **Paradermal section** (an oblique section running from upper epidermis to lower epidermis) of a **dicot** (lilac, *Syringa*) **leaf** (prep. slide, c.s., 18×). (Photo by J. W. Perry)

Figure 81e   Portion of **paradermal section** from Figure 81d (prep. slide, paradermal section, 140×). (Photo by J. W. Perry)

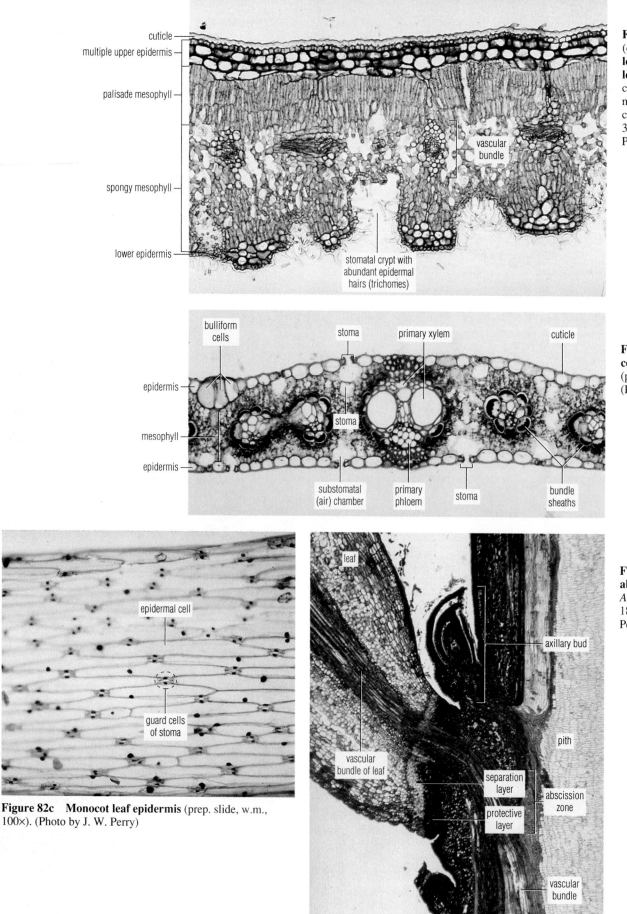

cuticle
multiple upper epidermis
palisade mesophyll
spongy mesophyll
lower epidermis
vascular bundle
stomatal crypt with abundant epidermal hairs (trichomes)

Figure 82a Dicot (oleander, *Nerium*) **leaf, a xeromorphic leaf**. Stomata are located in lower epidermis lining stomatal crypts (prep. slide, c.s., 30×). (Photo by J. W. Perry)

bulliform cells
stoma
primary xylem
cuticle
epidermis
mesophyll
epidermis
stoma
substomatal (air) chamber
primary phloem
stoma
bundle sheaths

Figure 82b Monocot (corn, *Zea*) **leaf** (prep. slide, c.s., 100×). (Photo by J. W. Perry)

epidermal cell
guard cells of stoma

Figure 82c Monocot leaf epidermis (prep. slide, w.m., 100×). (Photo by J. W. Perry)

leaf
axillary bud
pith
vascular bundle of leaf
separation layer
protective layer
abscission zone
vascular bundle

Figure 82d Leaf abscission (maple, *Acer*) (prep. slide, l.s., 18×). (Photo by J. W. Perry)

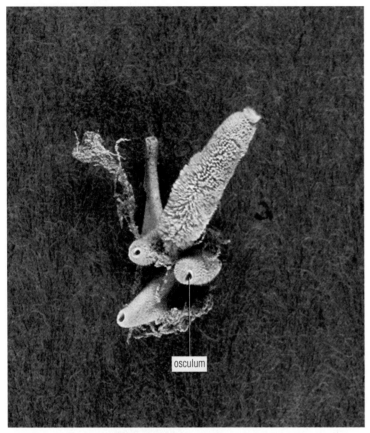

Figure 83a A cluster of dried specimens of the **syconoid sponge** *Scypha* (4×). (Photo by D. Morton)

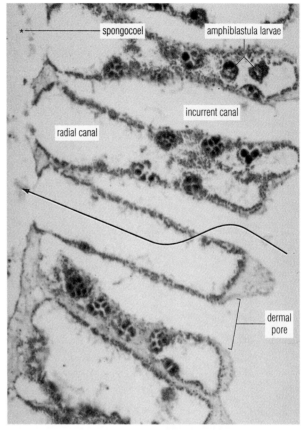

Figure 83b **Longitudinal section** of *Scypha*. The arrow indicates the path of water flow. Only a small part of the spongocoel can be seen (prep. slide, l.s., 120×). (Photo by D. Morton)

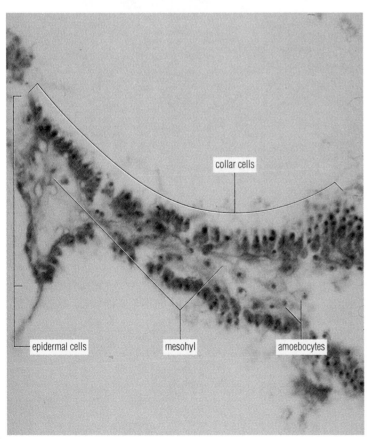

Figure 83c *Scypha* (prep. slide, l.s., 550×). (Photo by D. Morton)

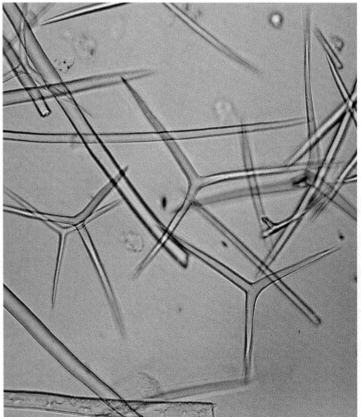

Figure 83d **Spicules of calcium carbonate** typical of the calcareous sponges like *Scypha* (prep. slide, w.m., 350×). (Photo by D. Morton)

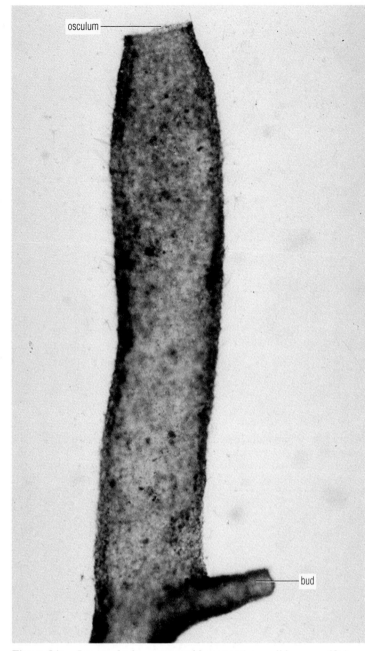

osculum

bud

Figure 84a *Leucosolenia*, an **asconoid sponge** (prep. slide, w.m., 40×). (Photo by D. Morton)

spicules

Figure 84b **Osculum** of *Leucosolenia* (prep. slide, w.m., 180×). (Photo by D. Morton)

Figure 84c **Bath sponge** (2/3×). (Photo by D. Morton)

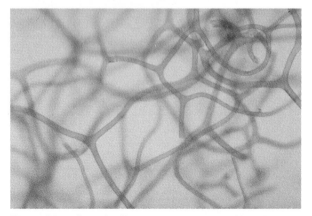

Figure 84d **Spongin fibers** (prep. slide, w.m., 80×). (Photo by D. Morton)

Figure 85a Dried **freshwater sponges**, *Spongilla fragilis*, with gemmules (1/2×). (Photo by W. A. Yoder)

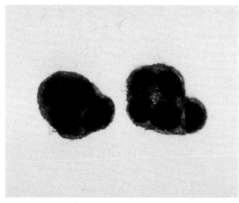

Figure 85b Resistant internal buds called **gemmules** formed by asexual reproduction in freshwater and some marine sponges (prep. slide, w.m., 30×). (Photo by D. Morton)

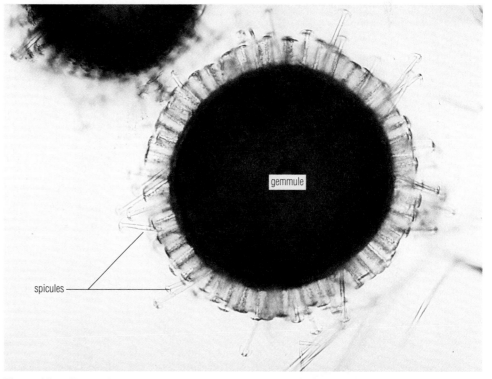

Figure 85c **Gemmule** (prep. slide, w.m., 150×). (Photo by D. Morton)

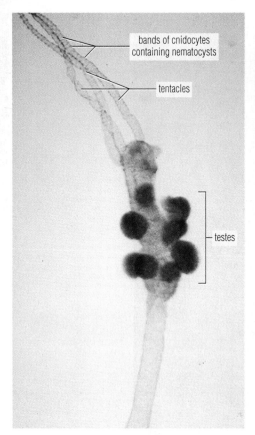

bands of cnidocytes
containing nematocysts

tentacles

testes

Figure 86a *Hydra* with **testes** (prep. slide, w.m., 30×). (Photo by D. Morton)

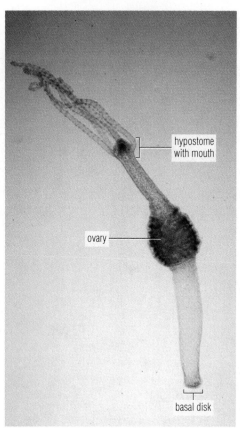

hypostome
with mouth

ovary

basal disk

Figure 86b *Hydra* with **ovary** (prep. slide, w.m., 30×). (Photo by D. Morton)

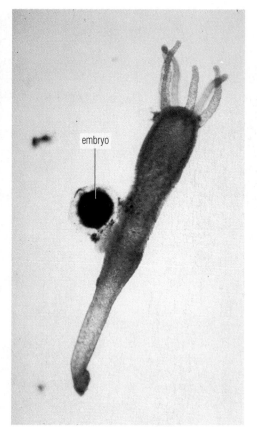

embryo

Figure 86c *Hydra* with **embryo** (prep. slide, w.m., 30×). (Photo by D. Morton)

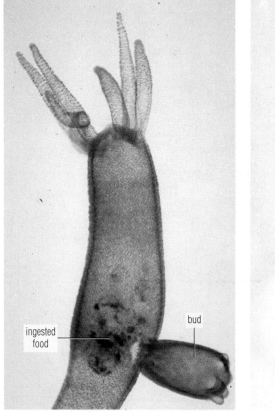

ingested
food

bud

Figure 86d *Hydra* with **bud** (prep. slide, w.m., 30×). (Photo by D. Morton)

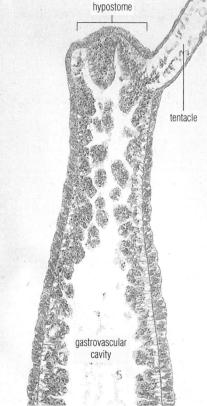

hypostome

tentacle

gastrovascular
cavity

Figure 86e **Longitudinal section** of *Hydra* (prep. slide, 100×). (Photo courtesy Biodisc, Inc.)

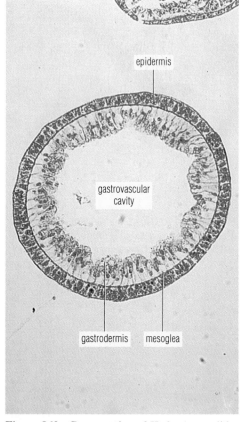

epidermis

gastrovascular
cavity

gastrodermis   mesoglea

Figure 86f **Cross section** of *Hydra* (prep. slide, 100×). (Photo courtesy Biodisc, Inc.)

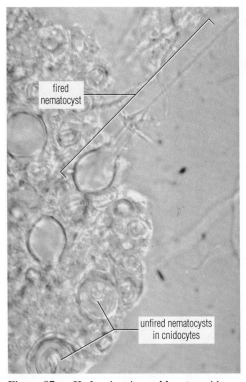

Figure 87a *Hydra* showing **cnidocytes** with fired and unfired nematocysts (live, w.m., 900×). (Photo by D. Morton)

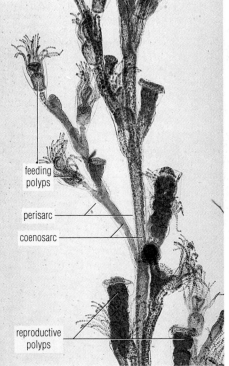

Figure 87b *Obelia* with **colonial polyps** (prep. slide, w.m., 30×). (Photo courtesy Biodisc, Inc.)

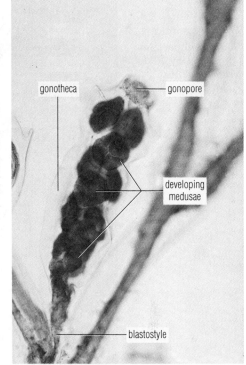

Figure 87c **Feeding polyp** of *Obelia* (prep. slide, w.m., 90×). (Photo by D. Morton)

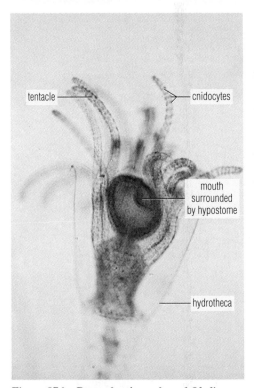

Figure 87d **Reproductive polyp** of *Obelia* (prep. slide, w.m., 150×). (Photo by D. Morton)

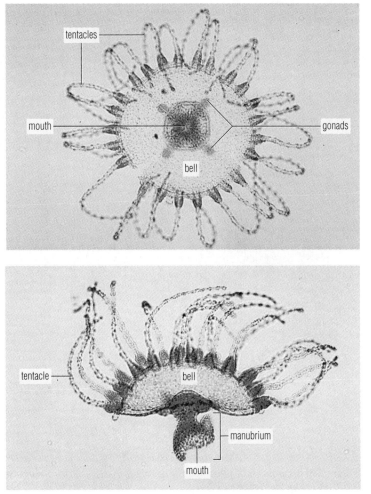

Figure 87e Top view of *Obelia* **medusa** (prep. slide, w.m., 120×). (Photo courtesy Biodisc, Inc.)

Figure 87f Side view of *Obelia* **medusa** (prep. slide, w.m., 150×). (Photo courtesy Biodisc, Inc.)

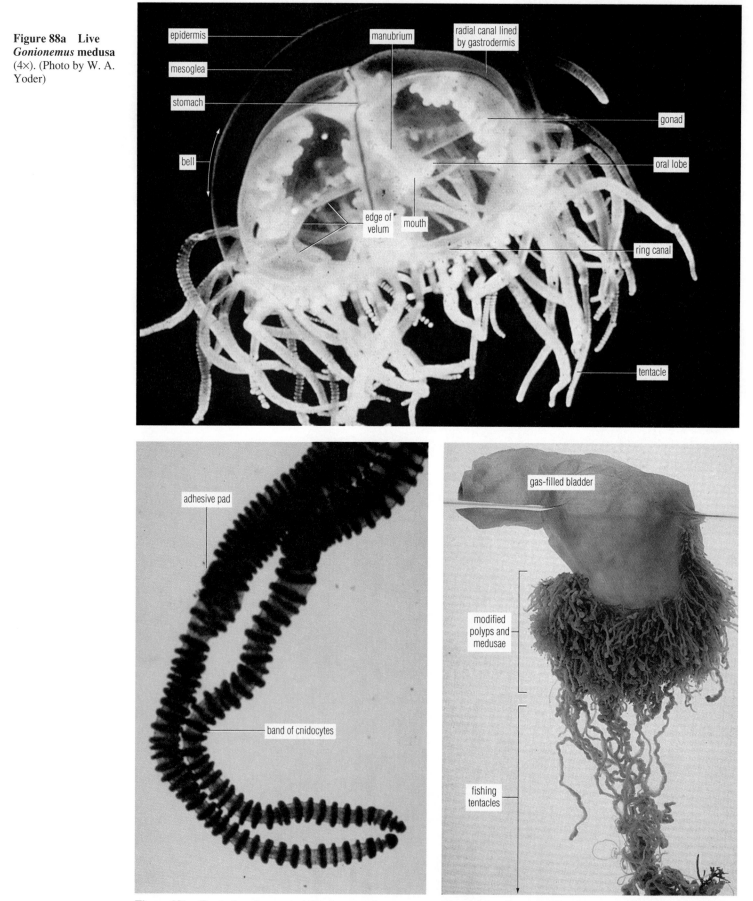

Figure 88a Live *Gonionemus* medusa (4×). (Photo by W. A. Yoder)

epidermis

mesoglea

stomach

bell

manubrium

radial canal lined by gastrodermis

gonad

oral lobe

edge of velum

mouth

ring canal

tentacle

Figure 88b **Tentacles** of preserved ***Gonionemus*** (w.m., 30×). (Photo by D. Morton)

adhesive pad

band of cnidocytes

Figure 88c Preserved **Portuguese man-of-war** with tentacles containing several types of polyps and medusae (1/3×). (Photo by D. Morton)

gas-filled bladder

modified polyps and medusae

fishing tentacles

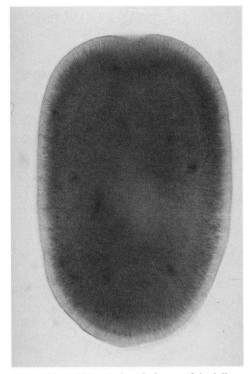

Figure 89a Ciliated **planula larva** of the jellyfish *Aurelia* (prep. slide, w.m., 400×). (Photo by D. Morton)

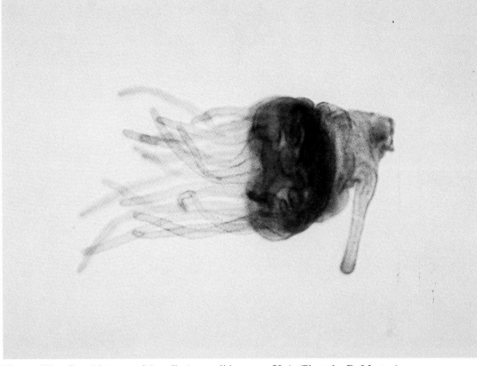

Figure 89b **Scyphistoma** of *Aurelia* (prep. slide, w.m., 50×). (Photo by D. Morton)

developing medusae (ephyrae)

Figure 89c **Strobila** of *Aurelia* (prep. slide, w.m., 30×). (Photo by D. Morton)

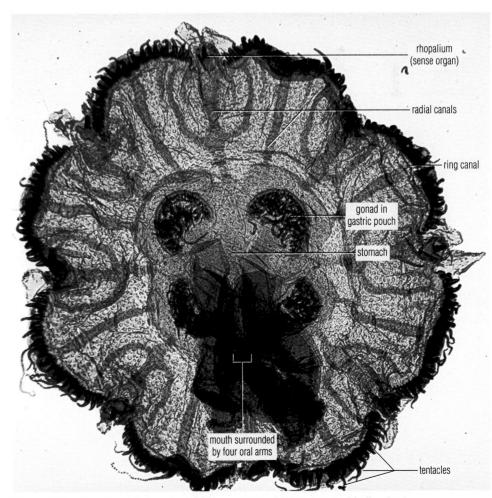

rhopalium (sense organ)

radial canals

ring canal

gonad in gastric pouch

stomach

mouth surrounded by four oral arms

tentacles

Figure 89d *Aurelia* **medusa** (prep. slide, w.m., 10×). (Photo courtesy Biodisc, Inc.)

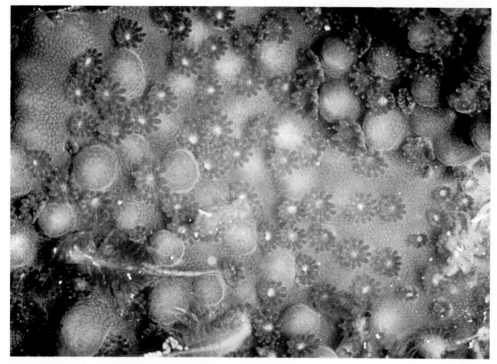

Figure 90a **Live coral polyps** (1.5×). (Photo by W. A. Yoder)

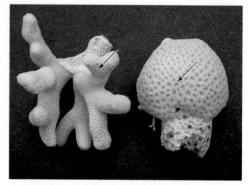

Figure 90b **Exoskeletons** of hard **coral**. The arrows indicate sites where individual polyps live (1/3×). (Photo by D. Morton)

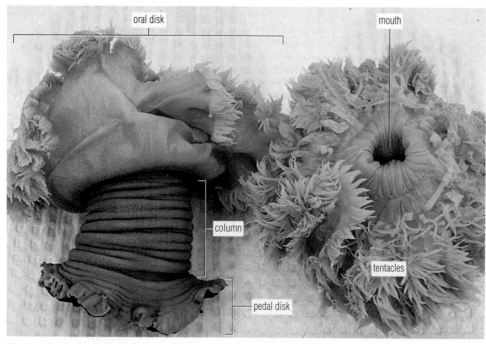

Figure 90c Top and side views of preserved **sea anemones** (1.5×). (Photo by D. Morton)

oral disk

mouth

column

tentacles

pedal disk

Figure 90d **Live sea anemones** (1/3×). (Photo by W. A. Yoder)

Figure 90e Pink-tipped **sea anemone**, *Condylactis gigantea*, **with mutualistic spotted cleaning shrimp**, *Periclimenes yucatanicus* (1×). (Photo by W. A. Yoder)

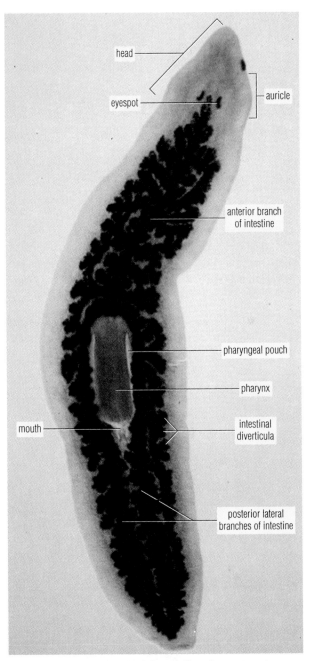

Figure 91a   **Planaria** with injected **digestive system** (prep. slide, w.m., 40×). (Photo by D. Morton)

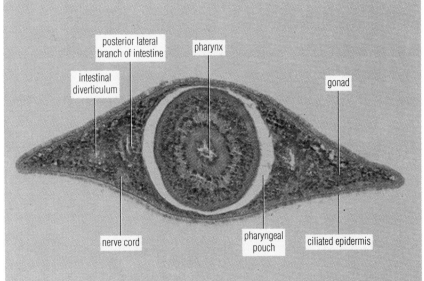

Figure 91b   Cross section of **planaria** (prep. slide, 150×). (Photo by D. Morton)

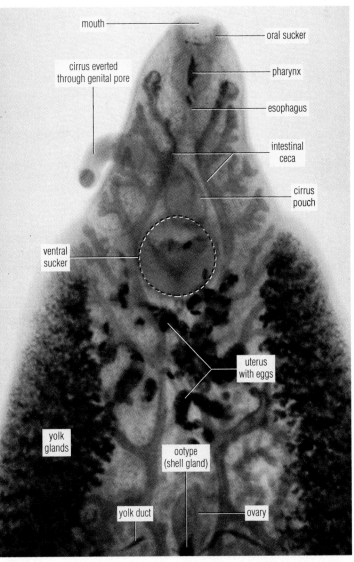

Figure 91c   Anterior end of the **sheep liver fluke**, *Fasciola hepatica*. The cirrus is an eversible copulatory organ (prep. slide, w.m., 20×). (Photo by D. Morton)

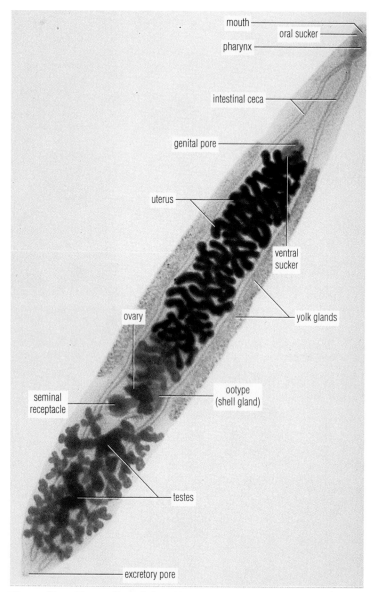

mouth
oral sucker
pharynx
intestinal ceca
genital pore
uterus
ventral sucker
ovary
yolk glands
seminal receptacle
ootype (shell gland)
testes
excretory pore

Figure 92a **Human liver fluke**, *Clonorchis sinensis* (prep. slide, w.m., 20×). (Photo by D. Morton)

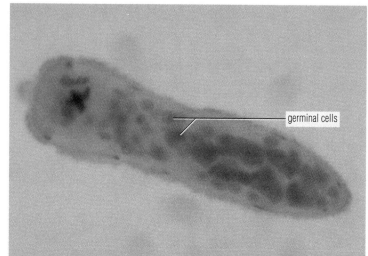

germinal cells

Figure 92b **Miracidium stage** of a **fluke** (prep. slide, w.m., 600×). (Photo by D. Morton)

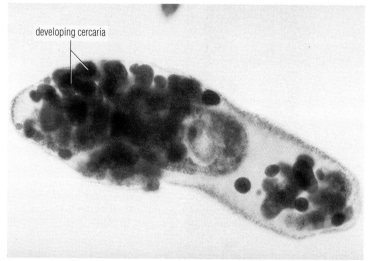

developing cercaria

Figure 92c **Redia stage** of a **fluke**, containing cercaria (prep. slide, w.m., 100×). (Photo by D. Morton)

Figure 92d **Cercaria stages** of **two different flukes** (prep. slide, w.m., 120×). (Photos by D. Morton)

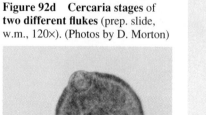

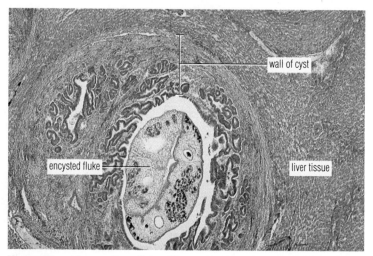

wall of cyst

encysted fluke

liver tissue

Figure 92e **Human liver fluke**, *Clonorchis sinensis*, **encysted in** a section of **human liver** (prep. slide, 40×). (Photo courtesy Biodisc, Inc.)

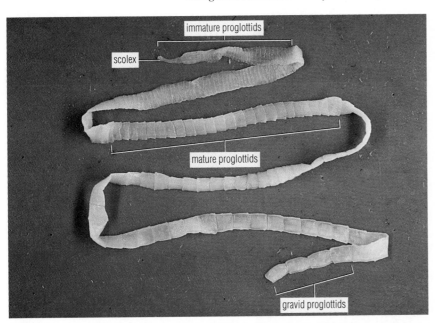

Figure 93a Preserved *Taenia*, a tapeworm (2/3×). (Photo by D. Morton)

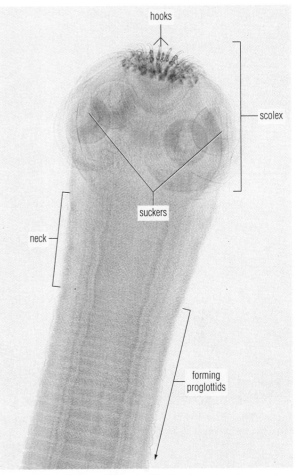

Figure 93b **Scolex** of *Taenia*. The hooks and suckers on the scolex are used to attach the tapeworm to the inside wall of the intestine. Nutrients are absorbed from the digested food (prep. slide, w.m., 40×). (Photo by D. Morton)

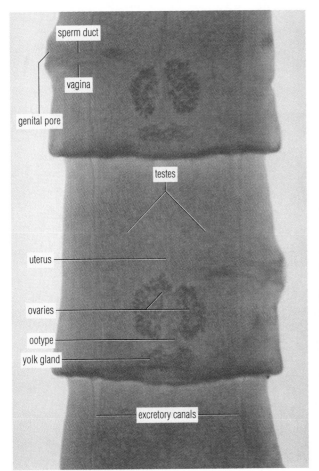

Figure 93c **Gravid proglottid** of *Taenia* (prep. slide, w.m., 30×). (Photo by D. Morton)

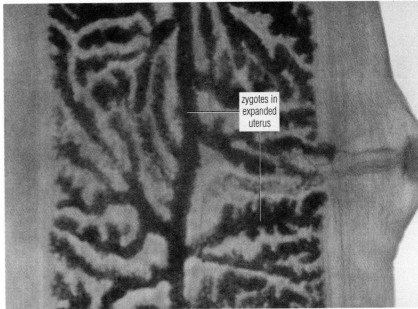

Figure 93d **Mature proglottid** of *Taenia* (prep. slide, w.m., 30×). (Photo by D. Morton)

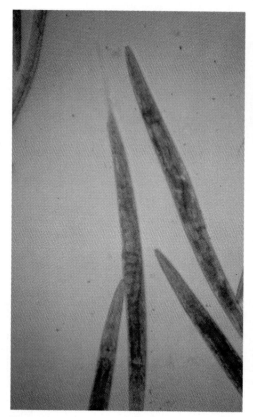

Figure 94a **Free-living nematodes**, *Rhabditella* females (live, w.m., 350×). (Photo by W. A. Yoder)

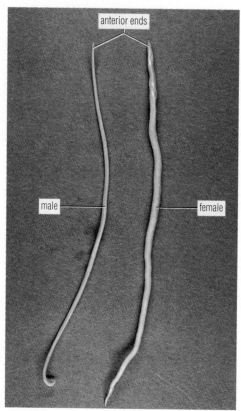

Figure 94b **Male and female** preserved specimens of *Ascaris lumbricoides*, a parasitic roundworm found in the human intestine (1/3×). (Photo by D. Morton)

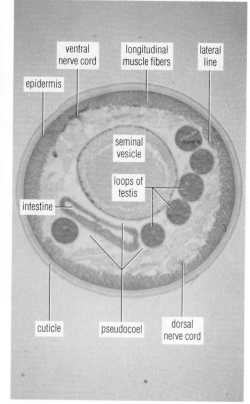

Figure 94c **Cross section** of male *Ascaris* (prep. slide, c.s., 40×). (Photo by D. Morton)

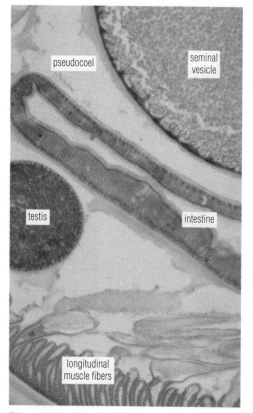

Figure 94d **Intestines** and **reproductive organs** of **male** *Ascaris* (prep. slide, c.s., 100×). (Photo by D. Morton)

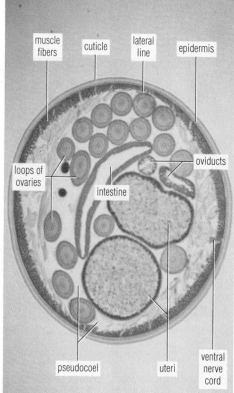

Figure 94e **Cross section** of female *Ascaris* (prep. slide, 30×). (Photo by D. Morton)

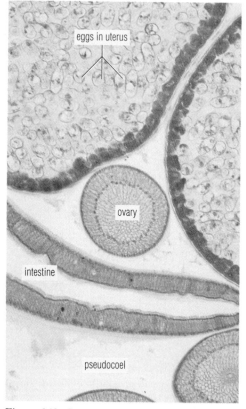

Figure 94f **Intestine** and **reproductive organs** of **female** *Ascaris* (prep. slide, c.s., 100×). (Photo by D. Morton)

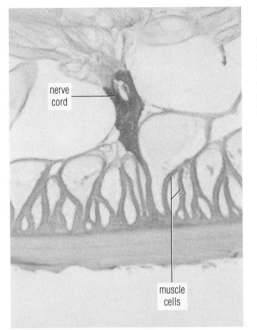

Figure 95a   Ventral nerve cord of *Ascaris* with attached muscle cells (prep. slide, c.s., 150×). (Photo by D. Morton)

nerve cord

muscle cells

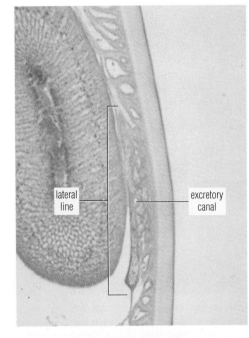

Figure 95b Lateral line and **excretory canal** of female *Ascaris* (prep. slide, c.s., 100×). (Photo by D. Morton)

lateral line

excretory canal

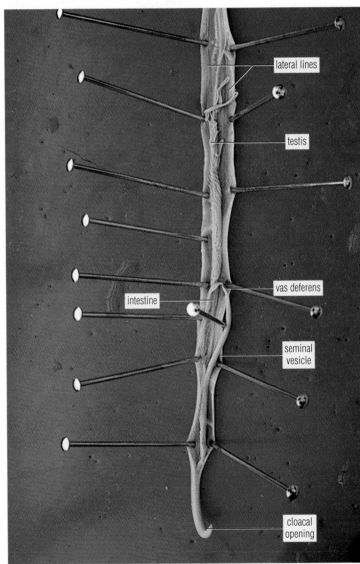

lateral lines

testis

vas deferens

intestine

seminal vesicle

cloacal opening

Figure 95c   Dissection of **male** *Ascaris lumbricoides* (1×). (Photo by D. Morton)

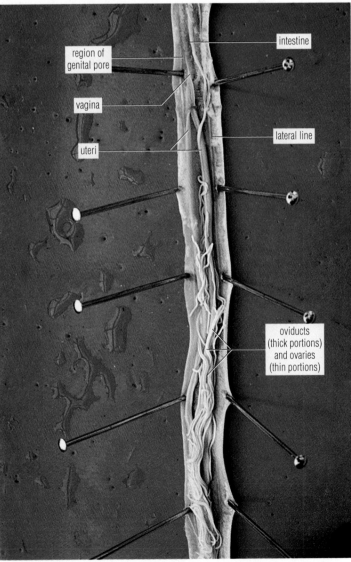

region of genital pore

intestine

vagina

uteri

lateral line

oviducts (thick portions) and ovaries (thin portions)

Figure 95d   Dissection of **female** *Ascaris lumbricoides* (1×) (Photo by D. Morton)

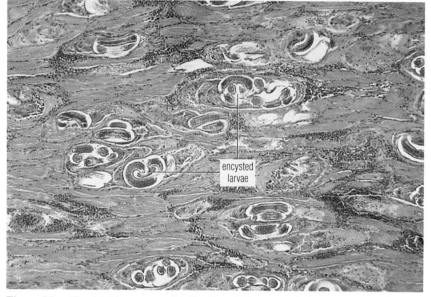

Figure 96a   The **trichina worm**, *Trichinella spiralis*, seen here encysted in a section of skeletal muscle, causes the disease trichinosis. In humans, this parasite is ingested by eating poorly cooked meat, usually pork (prep. slide, 70×). (Photo courtesy Biodisc, Inc.)

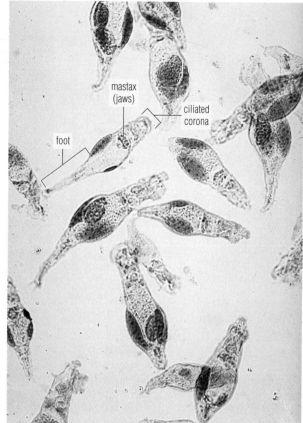

Figure 96b   **Rotifers** (prep. slide, w.m., 130×). (Photo courtesy Biodisc, Inc.)

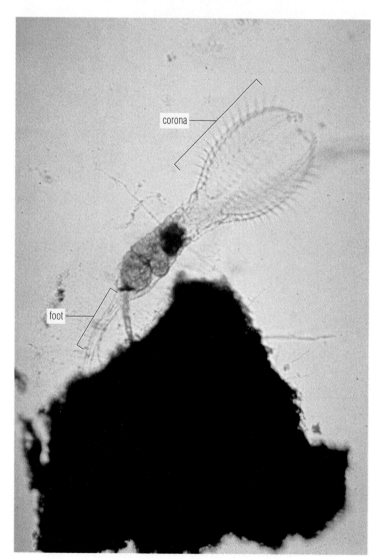

Figure 96c   **Live sessile rotifer**, *Stephanoceras* (w.m., 150×). (Photo by W. A. Yoder)

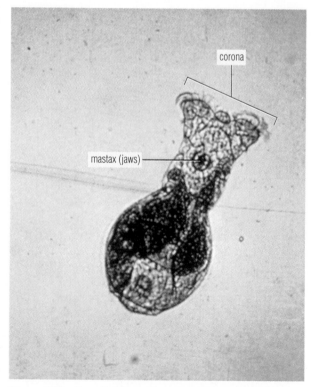

Figure 96d   **Free-living rotifer** (w.m., 350×). (Photo by W. A. Yoder)

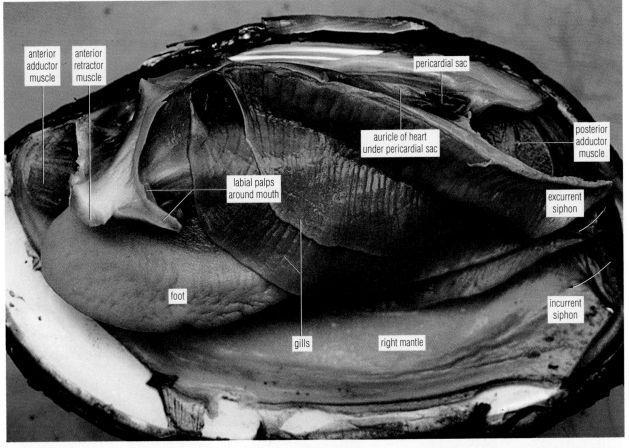

Figure 97a Dissection of freshwater clam. The left valve and mantle have been removed. Arrows indicate the path of water flow into the incurrent siphon and out the excurrent siphon (2×). (Photo by D. Morton)

anterior adductor muscle

anterior retractor muscle

pericardial sac

auricle of heart under pericardial sac

posterior adductor muscle

labial palps around mouth

excurrent siphon

foot

incurrent siphon

gills

right mantle

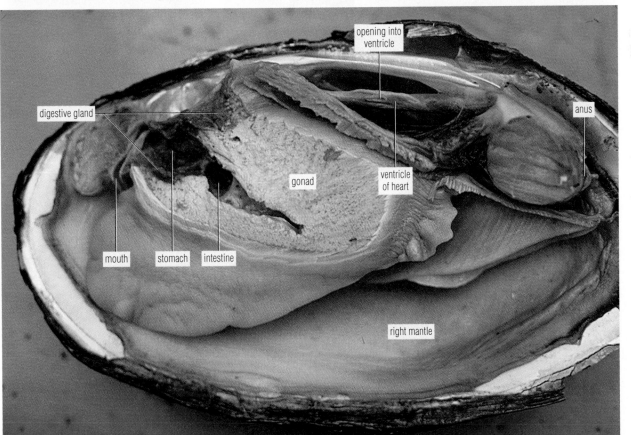

Figure 97b Further dissection of freshwater clam. The left gills and part of the visceral mass have been removed (2×). (Photo by D. Morton)

opening into ventricle

digestive gland

anus

gonad

ventricle of heart

mouth

stomach

intestine

right mantle

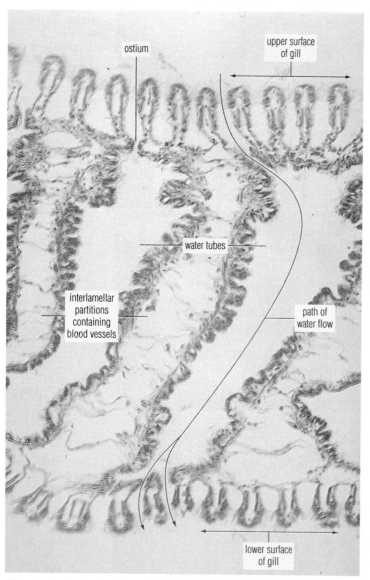

Figure 98a   **Clam gill** (prep. slide, c.s., 150×). (Photo courtesy Biodisc, Inc.)

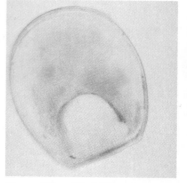

Figure 98b   **Glochidium larva** (prep. slide, w.m., 160×). (Photo by D. Morton)

Figure 98c   Freshwater **snail** (1×). (Photo by D. Morton)

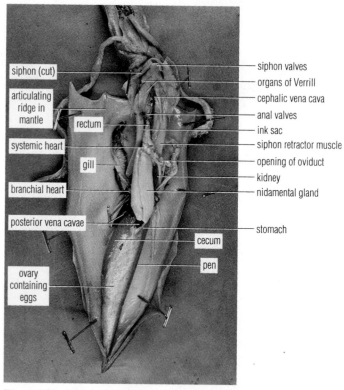

Figure 98f   **Internal anatomy** of female **squid** (1/3×). Nidamental glands and the ovary are absent in male specimens, and a penis is found in the same location as the oviduct. (Photo by D. Morton)

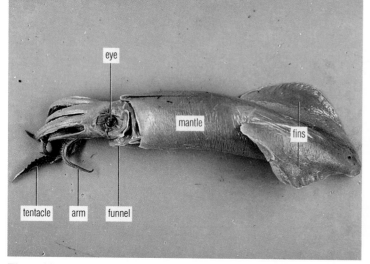

Figure 98d   **Lateral view** of a preserved **squid** (1/2×). (Photo by D. Morton)

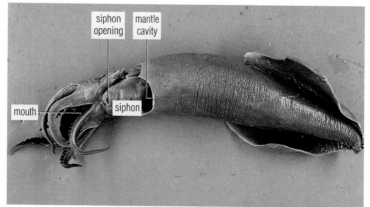

Figure 98e   **Ventral view** of a preserved **squid** (1/2×). (Photo by D. Morton)

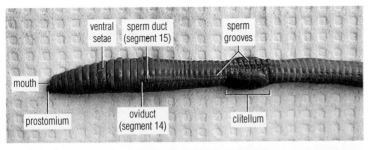

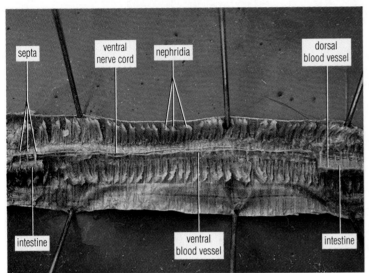

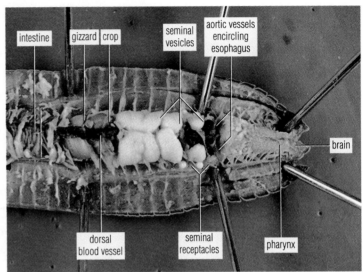

Figure 99a   **Ventral view** of preserved **earthworm** (1×). (Photo by D. Morton)

Figure 99b   **Dissection** of the **posterior** end of an **earthworm**. The intestine has been partially removed to better view the ventral nerve cord (2×). (Photo by D. Morton)

Figure 99c   **Dissection** of the **anterior** end of an **earthworm** (3×). (Photo by D. Morton)

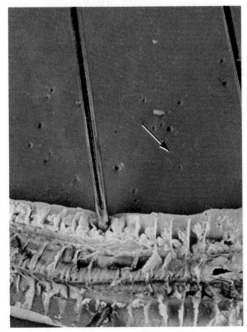

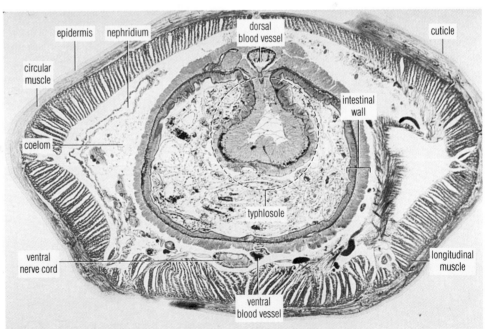

Figure 99d   **Brain** of an **earthworm** (arrow) (4×). (Photo by D. Morton)

Figure 99e   **Cross section** of **earthworm** (prep. slide, 20×). (Photo courtesy Biodisc, Inc.)

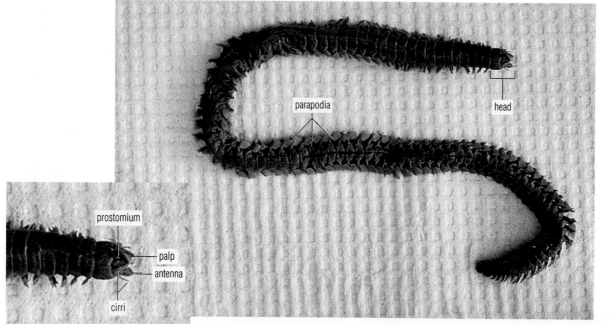

Figure 100a   Dorsal view of preserved **clamworm**, *Nereis* (1/2×). (Photo by D. Morton)

parapodia

head

prostomium

palp

antenna

cirri

Figure 100b Live clamworm, *Nereis* (1×). (Photo by W. A. Yoder)

dorsal blood vessel

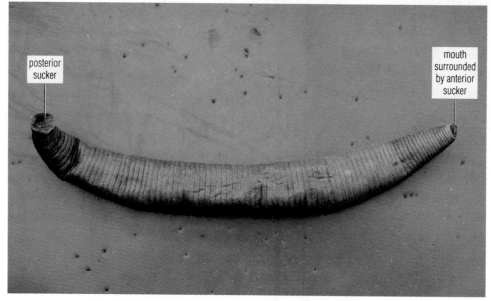

Figure 100c Preserved **leech** (2×). (Photo by D. Morton)

posterior sucker

mouth surrounded by anterior sucker

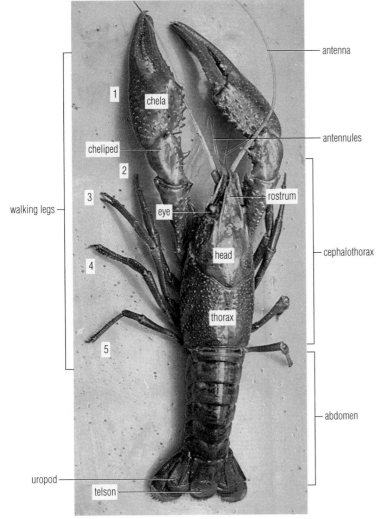

Figure 101a   **Dorsal view** of preserved **crayfish**. The five legs are numbered (2/3×). (Photo by D. Morton)

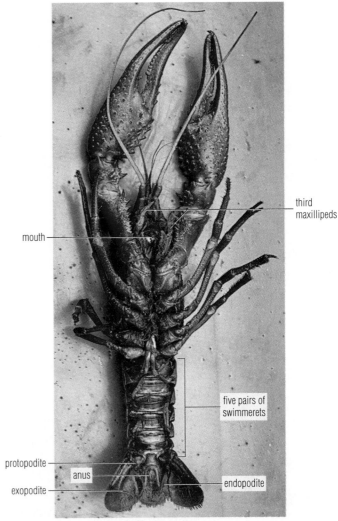

Figure 101b   **Ventral view** of preserved male **crayfish** (2/3×). (Photo by D. Morton)

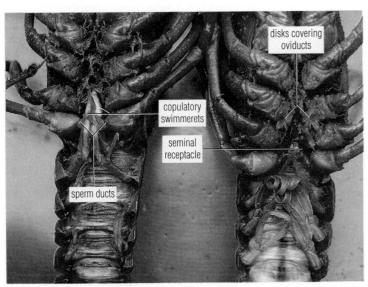

Figure 101c   **Ventral views** of **a male crayfish** (left) and **a female crayfish** (right) (1×). (Photo by D. Morton)

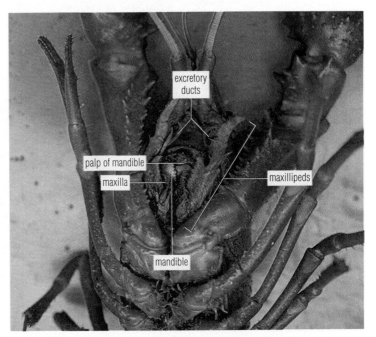

Figure 101d   **Ventral view** of **crayfish mouthparts** (2×). (Photo by D. Morton)

gills

Figure 102a
Gills of **crayfish.**
The lateral flap of
the carapace has been
removed (1×). (Photo
by D. Morton)

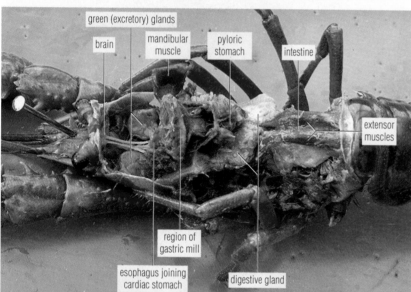

green (excretory) glands

brain

mandibular
muscle

pyloric
stomach

intestine

extensor
muscles

region of
gastric mill

esophagus joining
cardiac stomach

digestive gland

Figure 102b
Internal anatomy
of a **crayfish.** The
carapace, gills, and
heart have been re-
moved (1×). (Photo
by D. Morton)

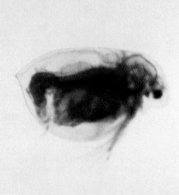

Figure 102c Preserved **water flea,**
Daphnia (prep. slide, w.m., 30×).
(Photo by D. Morton)

Figure 102d Live isopods, *Ligia exotica* (4×).
(Photo by W. A. Yoder)

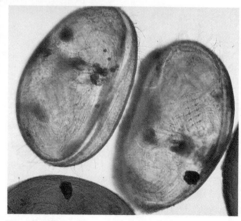

Figure 102e Ostracods (prep. slide, w.m.,
90×). (Photo by D. Morton)

Figure 102f Live acorn **barnacle** (1/2×).
(Photo by W. A. Yoder)

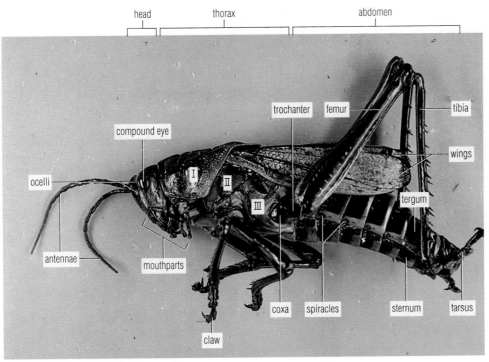

Figure 103a **External anatomy** of **grasshopper**. Roman numeral I indicates the prothorax, II indicates the mesothorax, and III indicates the metathorax (1×). (Photo by D. Morton)

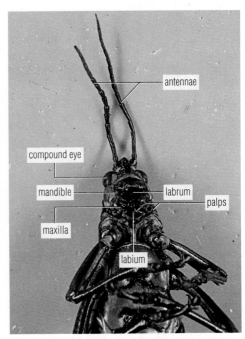

Figure 103c **Grasshopper mouthparts** (1×). (Photo by D. Morton)

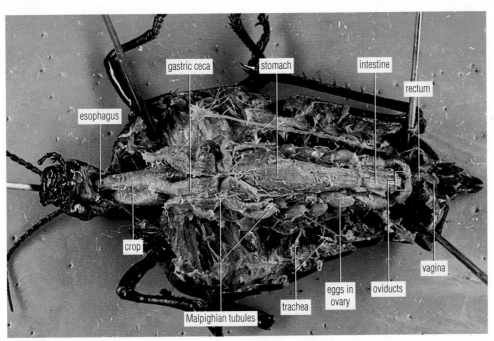

Figure 103b Ventral view of **internal anatomy** of **grasshopper** (1×). (Photo by D. Morton)

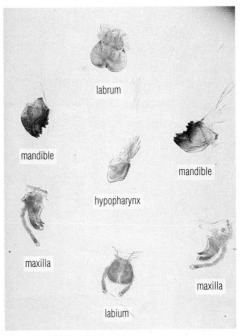

Figure 103d **Isolated grasshopper mouthparts** (prep. slide, w.m., 6×). (Photo courtesy Biodisc, Inc.)

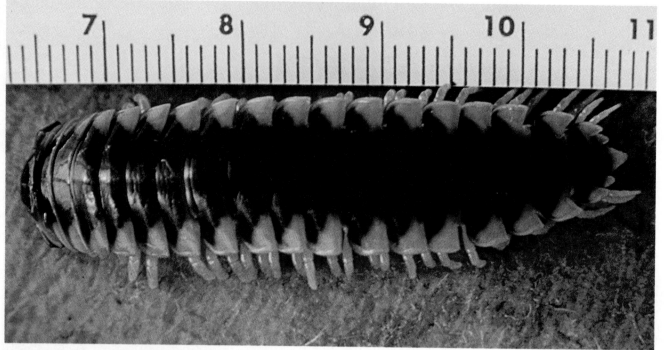

Figure 104a **Live millipede** (1.5×). (Photo by W. A. Yoder)

Figure 104b **Live centipede** eating lizard (1×). (Photo by W. A. Yoder)

Figure 104c **Live spider** and **web** (2×). (Photo by J. H. Howard)

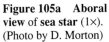

Figure 105a Aboral view of **sea star** (1×). (Photo by D. Morton)

Figure 105b Oral view of **sea star** (1×). (Photo by D. Morton)

Figure 105c Aboral view of **internal anatomy** of **sea star**. The digestive gland has been removed from one arm to show the gonad (3/4×). (Photo by D. Morton)

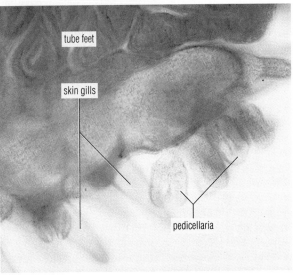

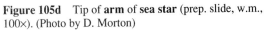

Figure 105d Tip of **arm** of **sea star** (prep. slide, w.m., 100×). (Photo by D. Morton)

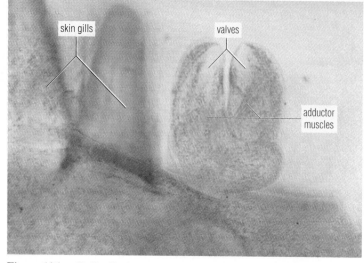

Figure 106a   Pedicellaria and **skin gills** of **sea star** (prep. slide, w.m., 250×). (Photo by D. Morton)

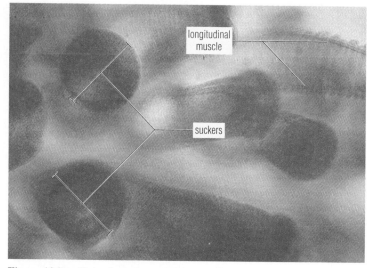

Figure 106b   Tube feet of **sea star** (prep. slide, w.m., 100×). (Photo by D. Morton)

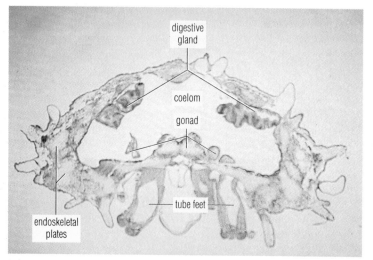

Figure 106c   Cross section of **sea star arm** (prep. slide, 30×). (Photo by D. Morton)

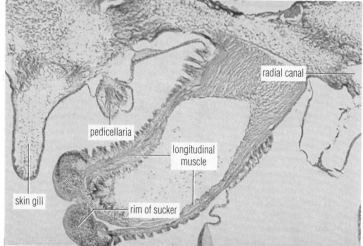

Figure 106d   Cross section of **tube foot** of **sea star** (prep. slide, 90×). (Photo by D. Morton)

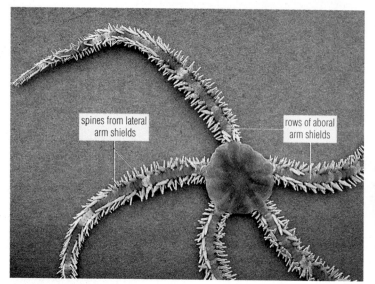

Figure 106e   Dried brittle star (2/3×). (Photo by D. Morton)

Figure 106f   Live reticulated **brittle star**, *Ophioneris reticulata* (1/3×). (Photo by W. A. Yoder)

Figure 107a Aboral (left) and oral (right) views of dried **sea urchins** (1/2×). (Photo by D. Morton)

periproct containing anus and surrounded by madreporite and genital pores

lip of peristomial membrane surrounding mouth

spines

teeth of lantern

Figure 107d Preserved **sea cucumber** (1/3×). (Photo by D. Morton)

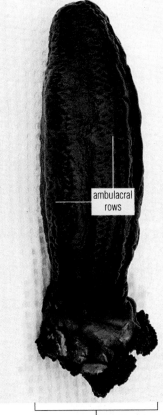

ambulacral rows

dendritic tentacles surrounding mouth

Figure 107b **Live** variegated **sea urchins**, *Lytechinus variegatus* (1/3×). (Photo by W. A. Yoder)

lunule

petaloid ambulacra

Figure 107c Dried **tests** of **sea biscuit** (left) and **sand dollar** (right) (1/2×). (Photo by D. Morton)

Figure 107e Live sea cucumber (1/3×). (Photo by W. A. Yoder)

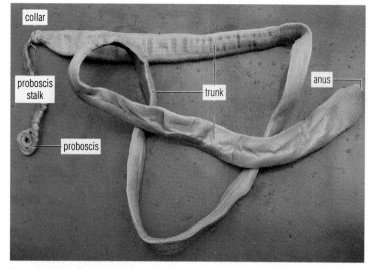

Figure 108a   Preserved **acorn worm** (1/2×). (Photo by D. Morton)

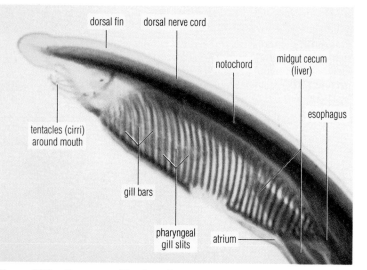

Figure 108b   **Low magnification** of anterior end of *Amphioxus* (prep. slide, w.m., 30×). (Photo by D. Morton)

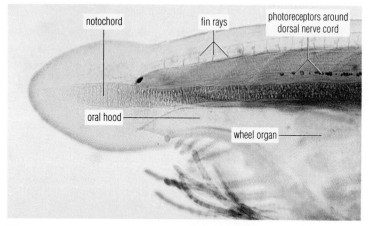

Figure 108c   **Anterior end** of *Amphioxus* (prep. slide, w.m., 100×). (Photo by D. Morton)

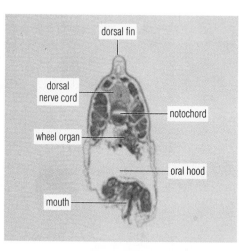

**Figure 108d
Anterior end** of *Amphioxus* (prep. slide, c.s., 30×). (Photo by D. Morton)

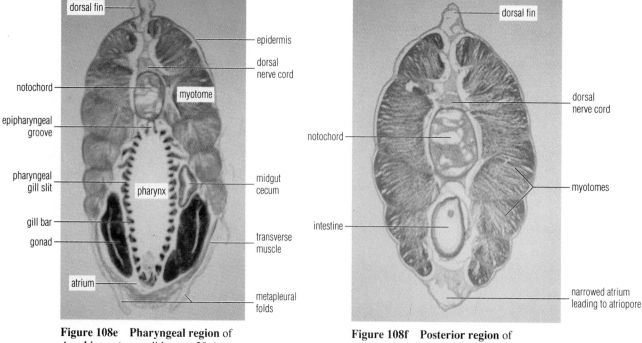

Figure 108e   **Pharyngeal region** of *Amphioxus* (prep. slide, c.s., 20×). (Photo by D. Morton)

Figure 108f   **Posterior region** of *Amphioxus* just anterior to anus (prep. slide, c.s., 20×). (Photo by D. Morton)

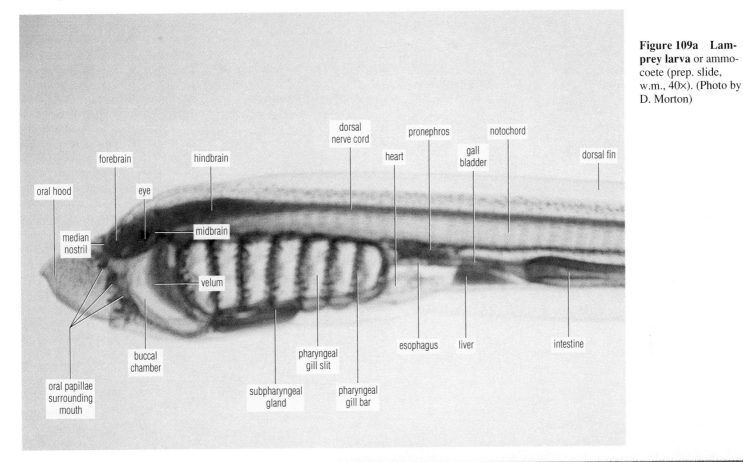

Figure 109a **Lamprey larva** or ammocoete (prep. slide, w.m., 40×). (Photo by D. Morton)

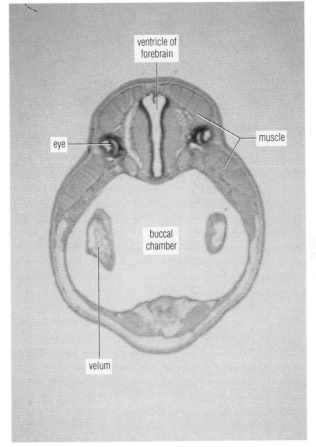

Figure 109b **Cross section** of anterior region of **lamprey larva** (prep. slide, 30×). (Photo by D. Morton)

Figure 109c Dried **shark jaw**. Note multiple rows of teeth (1/3×). (Photo by D. Morton)

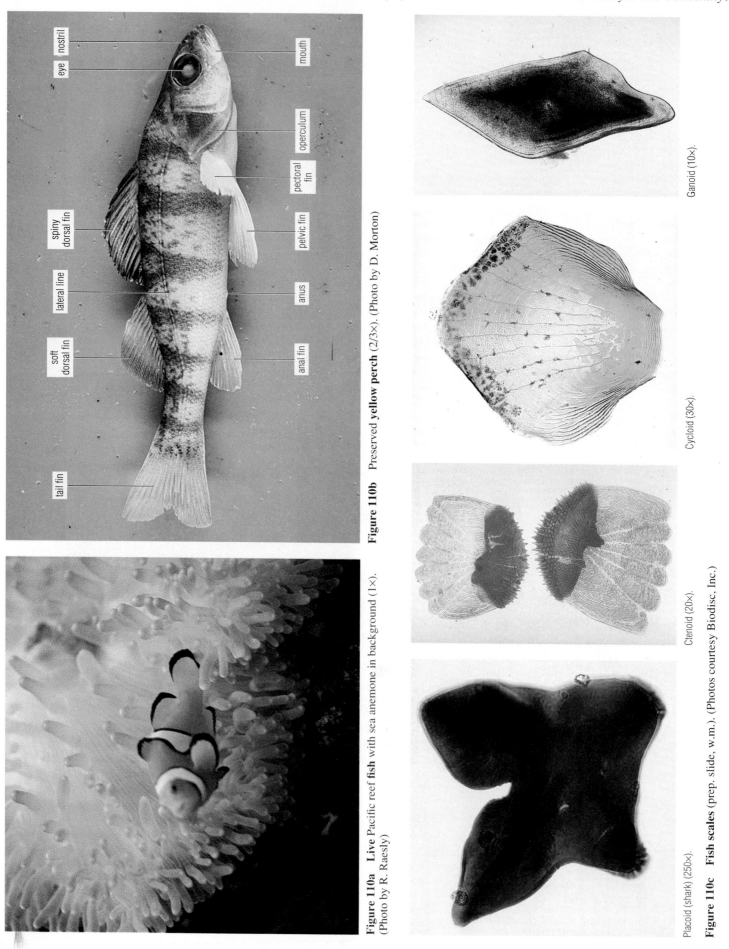

nostril

nostril

eye

mouth

operculum

pectoral fin

spiny dorsal fin

pelvic fin

lateral line

anus

soft dorsal fin

anal fin

tail fin

Figure 110b   Preserved **yellow perch** (2/3×). (Photo by D. Morton)

Ganoid (10×).

Cycloid (30×).

Ctenoid (20×).

Placoid (shark) (250×).

Figure 110a   **Live** Pacific reef **fish** with sea anemone in background (1×). (Photo by R. Raesly)

Figure 110c   **Fish scales** (prep. slide, w.m.). (Photos courtesy Biodisc, Inc.)

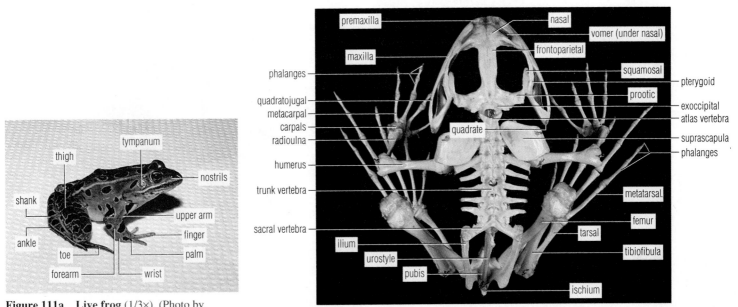

Figure 111a   Live frog (1/3×). (Photo by D. Morton)

Figure 111b   Bullfrog skeleton (1/2×). (Photo by D. Morton)

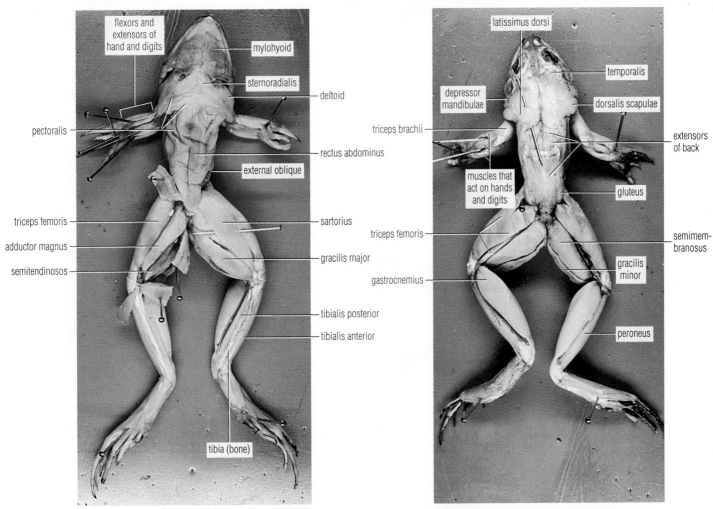

Figure 111c   Ventral view of **skeletal muscles** of preserved **frog** (2/3×). (Photo by D. Morton)

Figure 111d   Dorsal view of **frog skeletal muscles** (2/3×). (Photo by D. Morton)

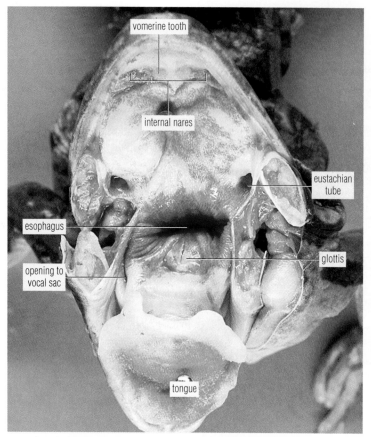

Figure 112a **Frog oral cavity** (2×). (Photo by D. Morton)

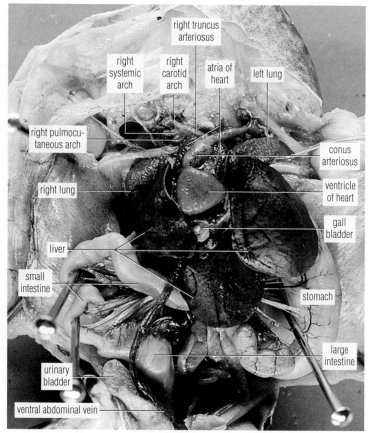

Figure 112b Ventral view of **internal anatomy** of male **frog** (2×). (Photo by D. Morton)

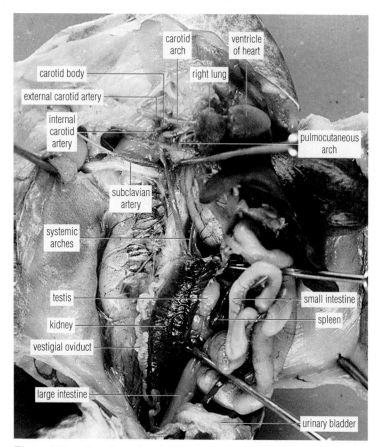

Figure 112c Male **frog**, showing **aortic arches** (2×). (Photo by D. Morton)

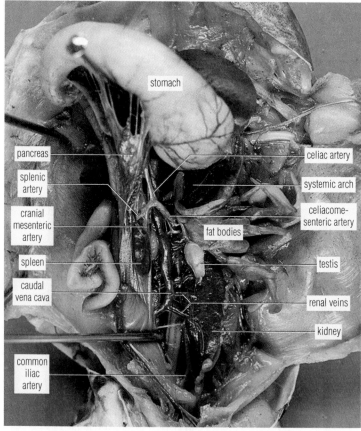

Figure 112d Male **frog**, showing **posterior vena cava**, **testis**, and **kidney** (2×). (Photo by D. Morton)

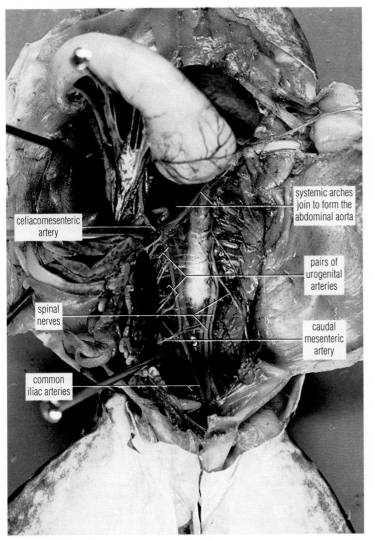

Figure 113a   Male **frog**, showing **abdominal aortic branches** (2×). (Photo by D. Morton)

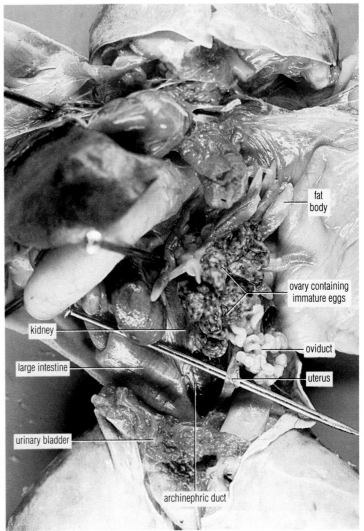

Figure 113b   Female **frog**, showing **ovary**, **oviduct**, and **uterus** (2×). (Photo by D. Morton)

Figure 113c   **Live** American **toads** in **amplexus** (1.5×). (Photo by W. A. Yoder)

Figure 113d   **Live** American **toad eggs** in pond water (2×). (Photo by W. A. Yoder)

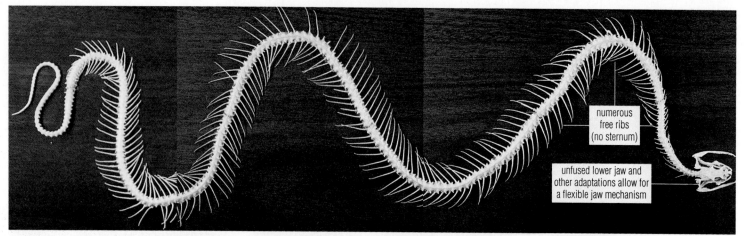

numerous free ribs (no sternum)

unfused lower jaw and other adaptations allow for a flexible jaw mechanism

Figure 114a **Snake skeleton** (1/2×). (Photos by D. Morton)

transparent scale over eye (no movable eyelids)

epidermal scales

Figure 114b Preserved **snake**. Snakes are further characterized by the absence of a tympanic membrane and absent or vestigial girdles (1/2×). (Photo by D. Morton)

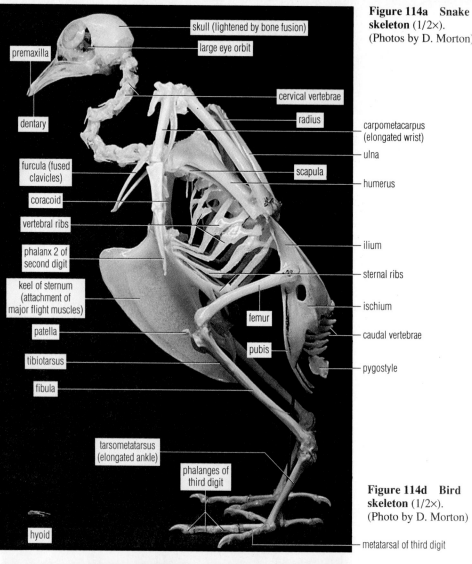

skull (lightened by bone fusion)

large eye orbit

premaxilla

cervical vertebrae

radius

carpometacarpus (elongated wrist)

ulna

dentary

scapula

humerus

furcula (fused clavicles)

coracoid

vertebral ribs

ilium

sternal ribs

phalanx 2 of second digit

keel of sternum (attachment of major flight muscles)

ischium

femur

caudal vertebrae

patella

pubis

pygostyle

tibiotarsus

fibula

tarsometatarsus (elongated ankle)

phalanges of third digit

Figure 114d **Bird skeleton** (1/2×). (Photo by D. Morton)

hyoid

metatarsal of third digit

Figure 114c **Live birds** (*Anas sibilatrix*, the Chiloe Wigeon) in flight (1/20×). (Photo by G. L. Brewer)

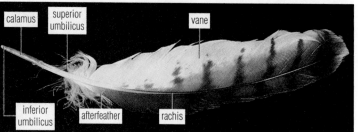

calamus

superior umbilicus

vane

inferior umbilicus

afterfeather

rachis

Figure 114e **Feather** (1/2×). (Photo by D. Morton)

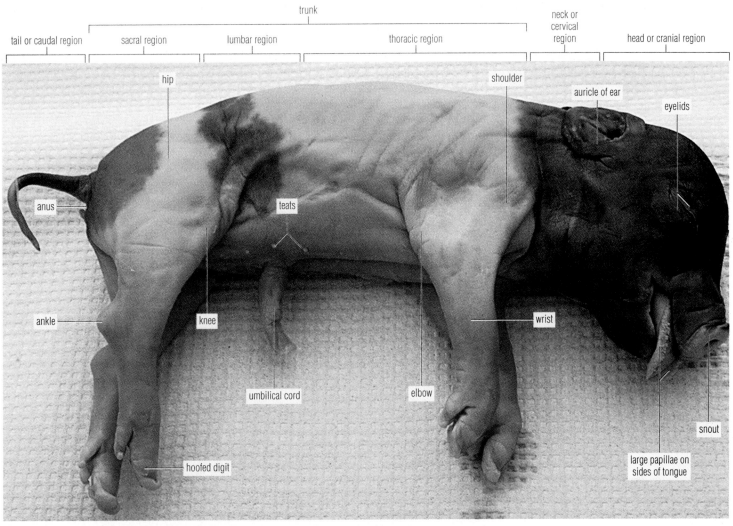

trunk

tail or caudal region | sacral region | lumbar region | thoracic region | neck or cervical region | head or cranial region

hip

shoulder

auricle of ear

eyelids

anus

teats

knee

ankle

wrist

umbilical cord

elbow

snout

hoofed digit

large papillae on sides of tongue

Figure 115a   External anatomy of a preserved **fetal pig** (2/3×). (Photo by D. Morton)

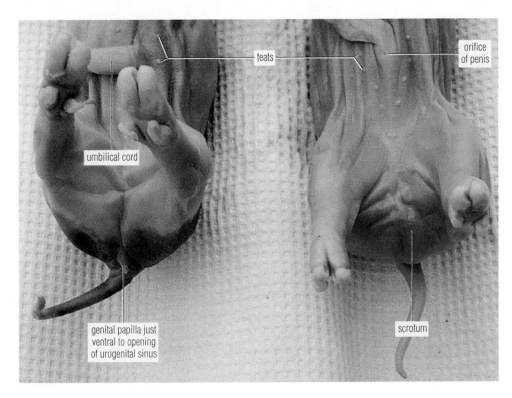

teats

orifice of penis

umbilical cord

genital papilla just ventral to opening of urogenital sinus

scrotum

Figure 115b   Ventral view of **female** (left) and **male** (right) **fetal pigs** (1/2×). (Photo by D. Morton)

Figure 116a   Ventral view of **internal anatomy of fetal pig** (1.25×). (Photo by D. Morton)

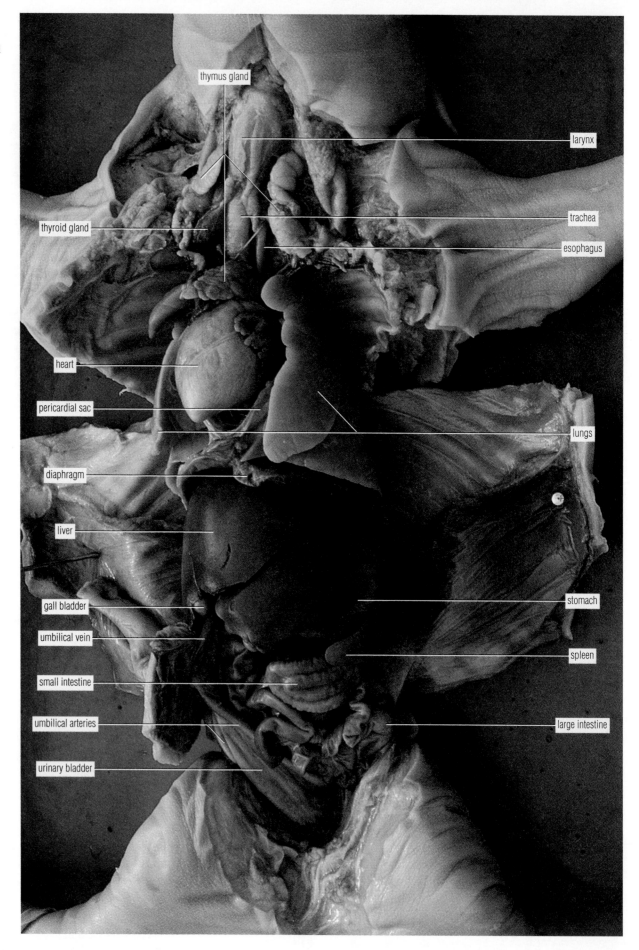

thymus gland

larynx

thyroid gland

trachea

esophagus

heart

pericardial sac

lungs

diaphragm

liver

gall bladder

stomach

umbilical vein

spleen

small intestine

umbilical arteries

large intestine

urinary bladder

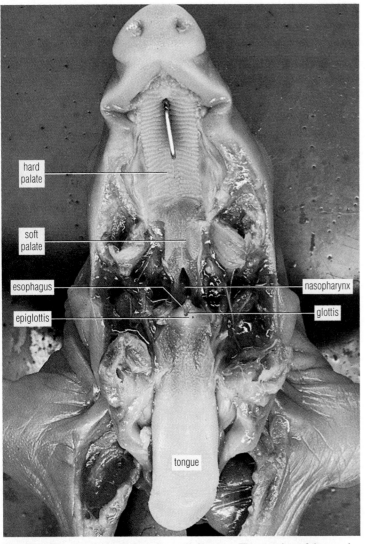

Figure 117a   **Fetal pig oral cavity** and **pharynx**. The opening of the esophagus can't be seen but is just behind the glottis (1×). (Photo by D. Morton)

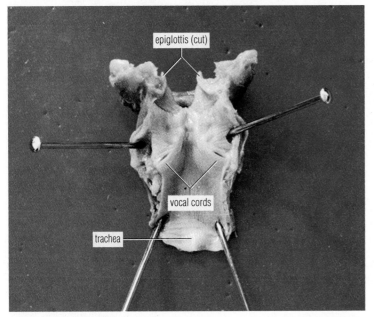

Figure 117b   Ventral view of **fetal pig larynx**, which has been slit open to show the vocal cords (2.5×). (Photo by D. Morton)

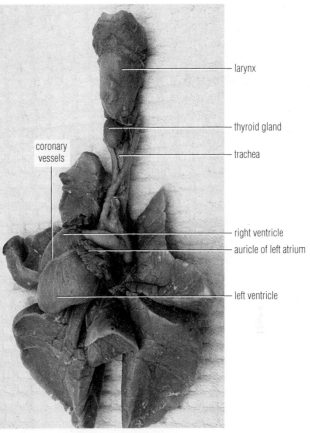

Figure 117c   **Ventral view** of **respiratory system** of **fetal pig** (1×). (Photo by D. Morton)

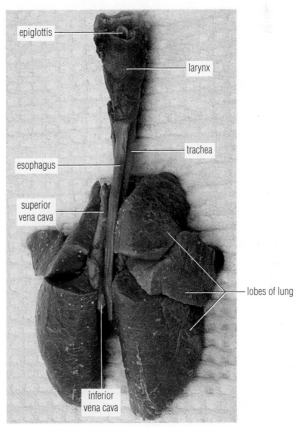

Figure 117d   **Dorsal view** of **respiratory system** of **fetal pig** (1×). (Photo by D. Morton)

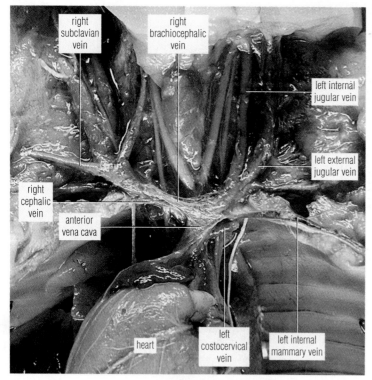

Figure 118a Ventral view of **thoracic veins** of **fetal pig** (2.5×). (Photo by D. Morton)

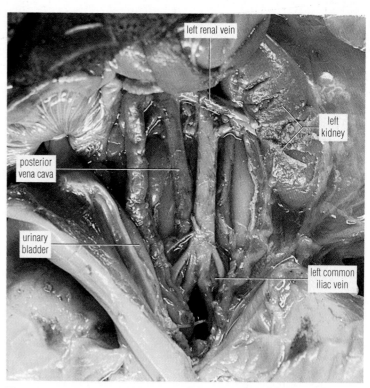

Figure 118b Ventral view of **abdominopelvic veins** of **fetal pig** (2.5×). (Photo by D. Morton)

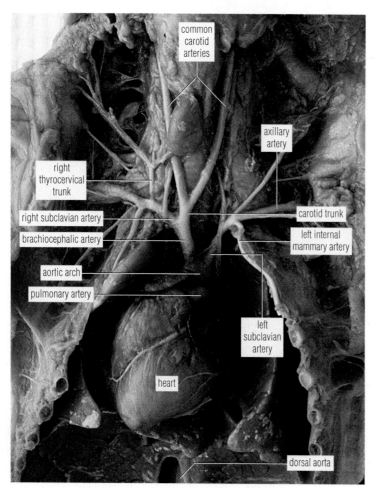

Figure 118c Ventral view of **thoracic arteries** of **fetal pig** (2×). (Photo by D. Morton)

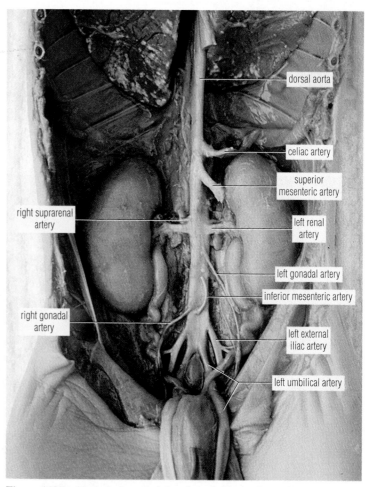

Figure 118d Ventral view of **abdominopelvic arteries** of **fetal pig** (2×). (Photo by D. Morton)

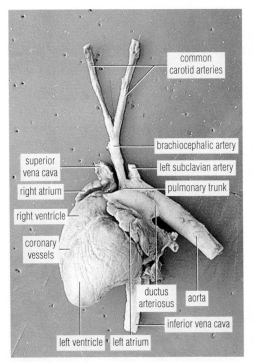

Figure 119a Ventral view of **fetal pig heart** (1.25×). (Photo by D. Morton)

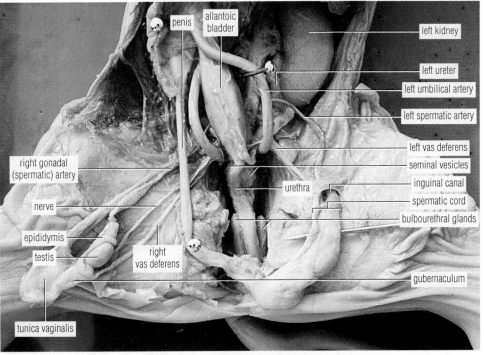

Figure 119b Ventral view of the **urogenital system** of the **male fetal pig** (1×). (Photo by D. Morton)

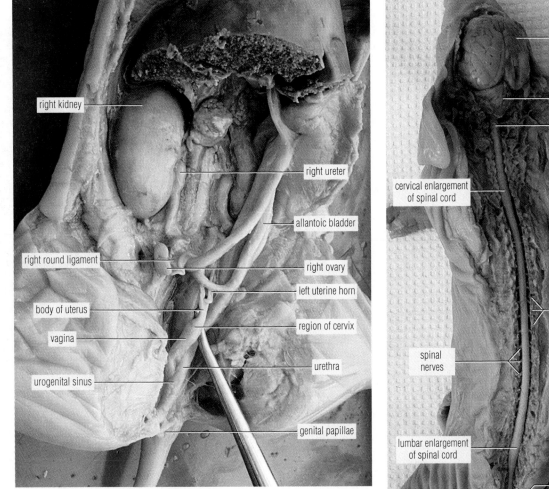

Figure 119c Ventral view of the **urogenital system** of the **female fetal pig** (1×). (Photo by D. Morton)

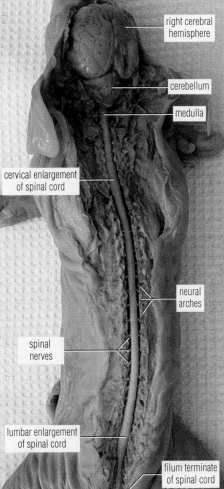

Figure 119d
Dorsal view of **central nervous system** of **fetal pig** (1/2×). (Photo by D. Morton)

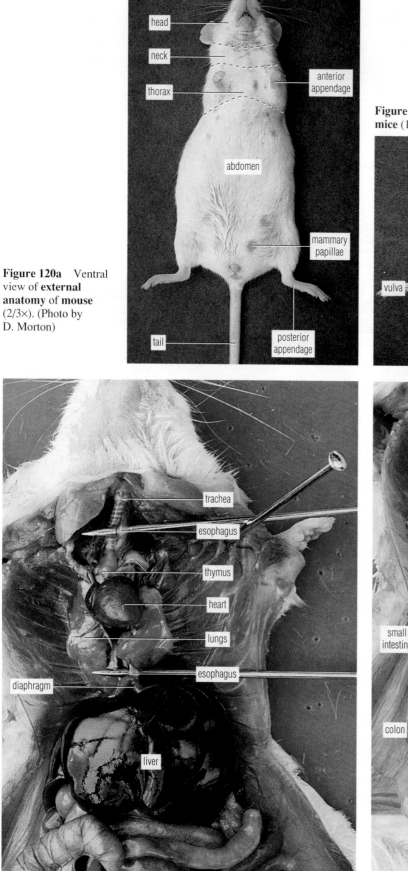

Figure 120a Ventral view of **external anatomy** of **mouse** (2/3×). (Photo by D. Morton)

Figure 120b Ventral view of **female** (left) and **male** (right) **mice** (1×). (Photo by D. Morton)

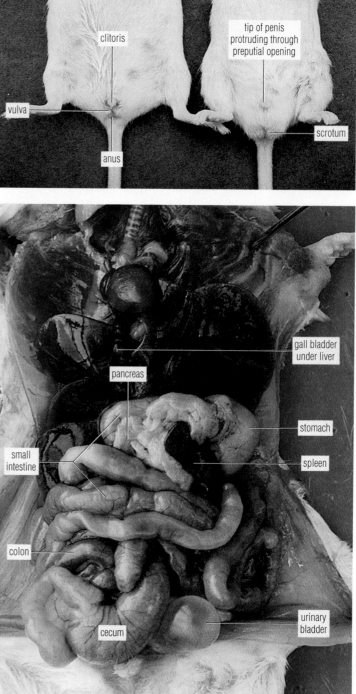

Figure 120c Ventral view of internal organs in **thoracic cavity** of recently sacrificed **mouse** (2×). (Photo by D. Morton)

Figure 120d Ventral view of internal organs in **abdominopelvic cavity** of **mouse** (2×). (Photo by D. Morton)

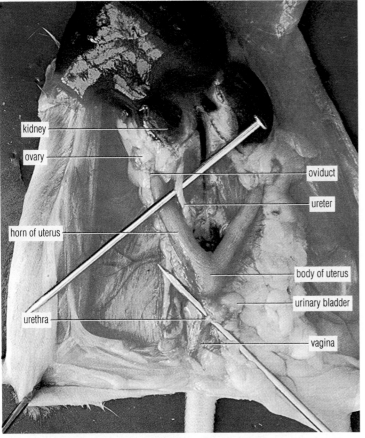

Figure 121a   Ventral view of the **urogenital system** of the **female mouse** (2×). (Photo by D. Morton)

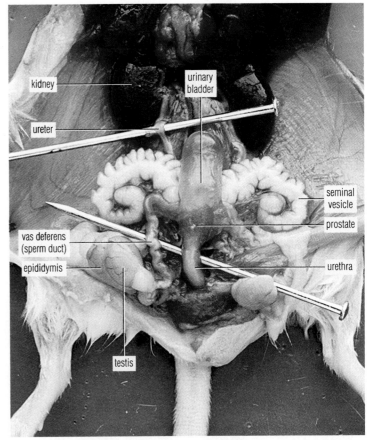

Figure 121b   Ventral view of the **urogenital system** of the **male mouse** (2×). (Photo by D. Morton)

Figure 121c   Female dog with **nursing puppies**. Hair and feeding milk from mammary glands are two characteristics of mammals (1/10×). (Photo by D. Morton)

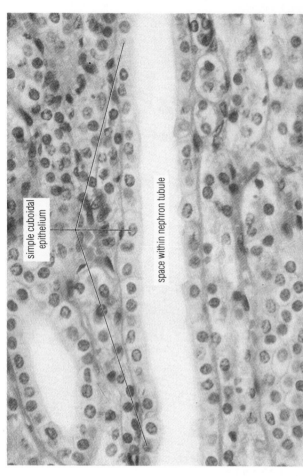

Figure 122a **Simple squamous epithelium** (section of outer layer of Bowman's capsule) in and **simple cuboidal epithelium** (cross section of the tubular portion of a nephron) in cortex of kidney (prep. slide, 500×). (Photo by D. Morton)

simple squamous epithelium

space within nephron tubule

glomerulus

space within Bowman's capsule

simple cuboidal epithelium

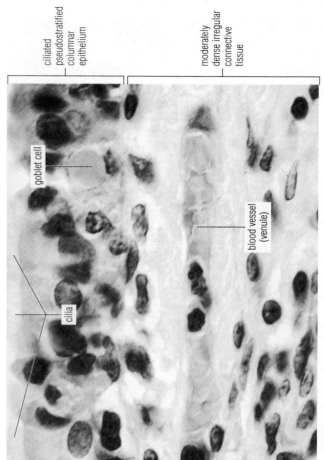

Figure 122b **Simple cuboidal epithelium** (longitudinal section of a nephron tubule) in kidney (prep. slide, 500×). (Photo by D. Morton)

simple cuboidal epithelium

space within nephron tubule

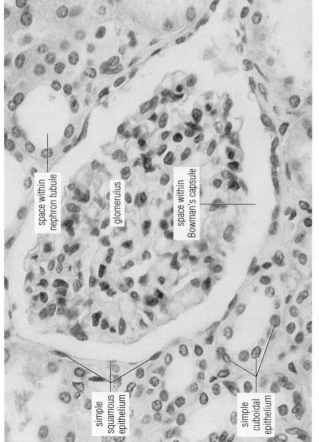

Figure 122c **Simple columnar epithelium** in a section of the small intestine (prep. slide, 500×). (Photo by D. Morton)

goblet cells

simple columnar epithelium

brush border

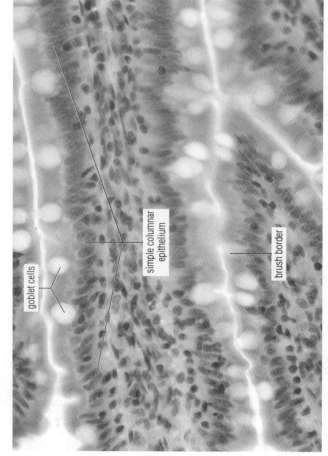

Figure 122d **Ciliated pseudostratified columnar epithelium** in a section of the trachea (prep. slide, 1200×). (Photo by D. Morton)

ciliated pseudostratified columnar epithelium

moderately dense irregular connective tissue

goblet cell

cilia

blood vessel (venule)

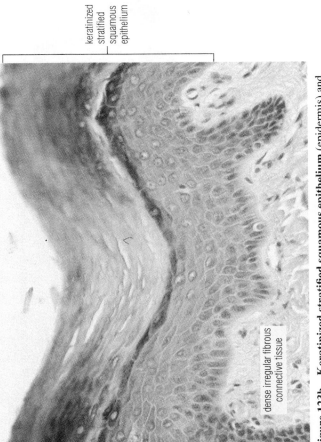

keratinized
stratified
squamous
epithelium

dense irregular fibrous
connective tissue

Figure 123b **Keratinized stratified squamous epithelium** (epidermis) and **dense irregular fibrous connective tissue** (dermis) in a section of skin (prep. slide, 150×). (Photo by D. Morton)

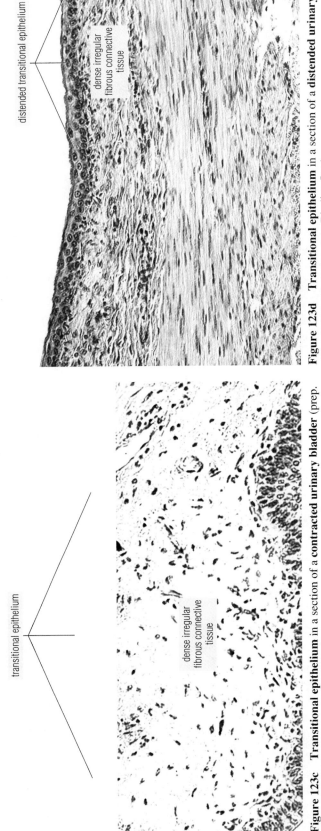

distended transitional epithelium

dense irregular
fibrous connective
tissue

Figure 123d **Transitional epithelium** in a section of a **distended urinary bladder** (prep. slide, 180×). (Photo courtesy Biodisc, Inc.)

dense irregular
fibrous connective
tissue

stratified
squamous
epithelium

Figure 123a **Stratified squamous epithelium** in a section of the esophagus (prep. slide, 700×). (Photo by D. Morton)

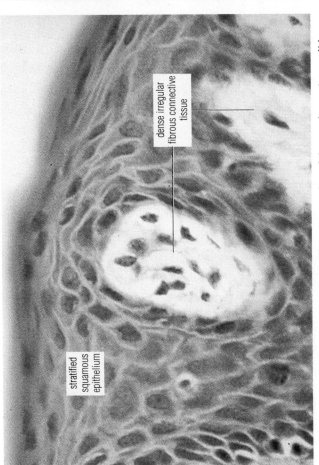

transitional epithelium

dense irregular
fibrous connective
tissue

Figure 123c **Transitional epithelium** in a section of a **contracted urinary bladder** (prep. slide, 220×). (Photo courtesy Biodisc, Inc.)

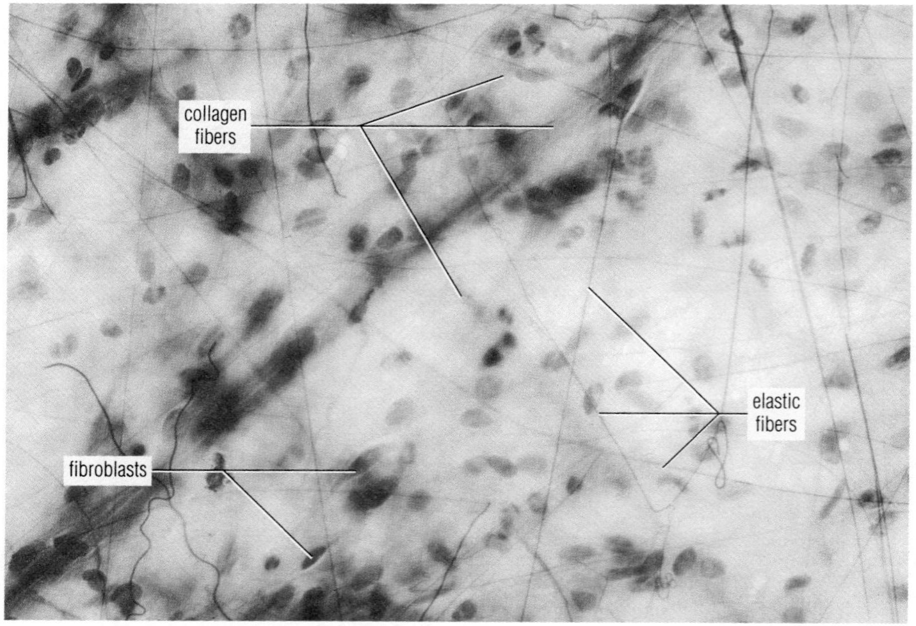

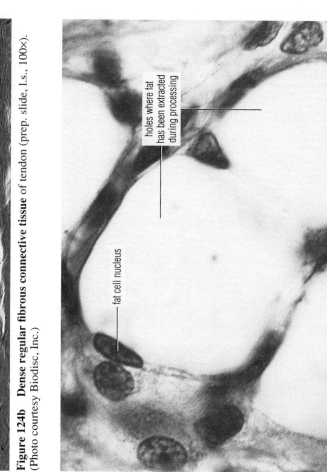

rows of fibroblasts between collagen fibers

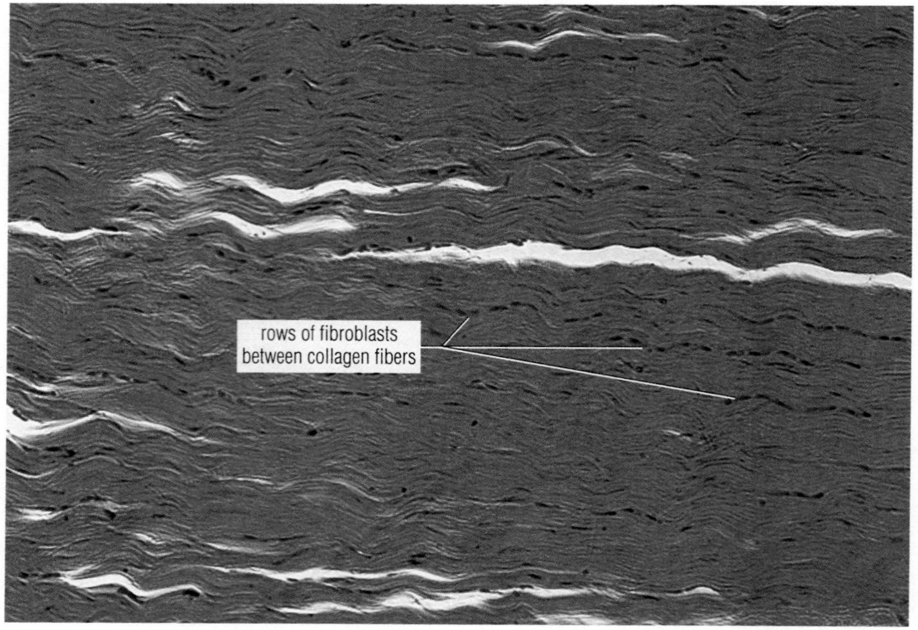

Figure 124b   Dense regular fibrous connective tissue of tendon (prep. slide, l.s., 100×). (Photo courtesy Biodisc, Inc.)

holes where fat has been extracted during processing

fat cell nucleus

Figure 124d   Section of unilocular **fat cells** or adipocytes (prep. slide, 1300×). (Photo by D. Morton)

elastic fibers

collagen fibers

fibroblasts

Figure 124a   Loose (areolar) **connective tissue** of mesentery (prep. slide, w.m., 500×). (Photo by D. Morton)

fat cells

capillaries

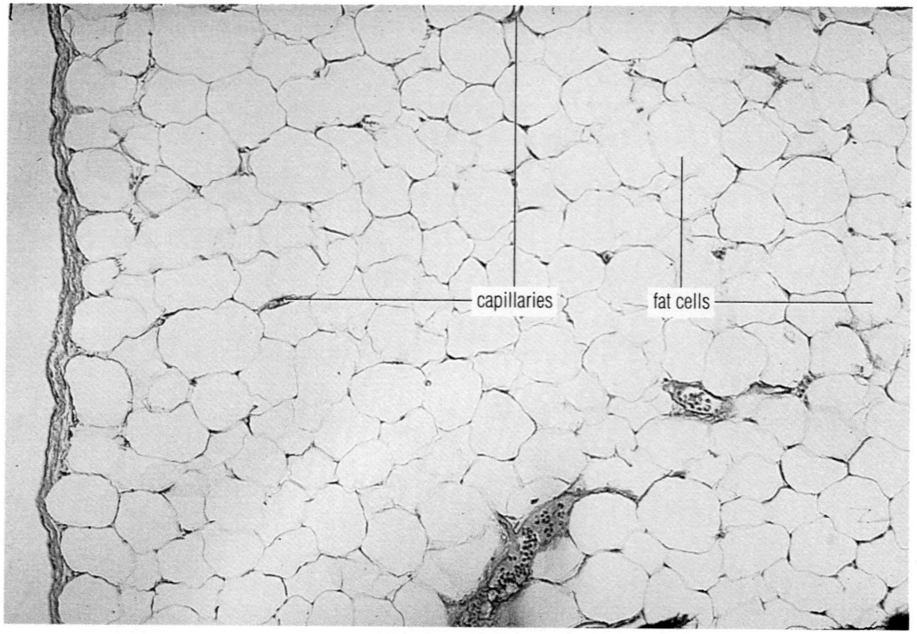

Figure 124c   Section of **adipose connective tissue** (prep. slide, 150×). (Photo courtesy Biodisc, Inc.)

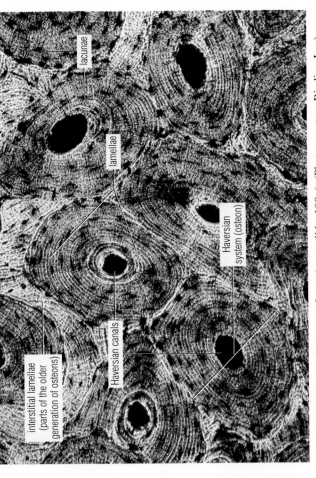

Figure 125b   Ground **compact bone** (prep. slide, 100×). (Photo courtesy Biodisc, Inc.)

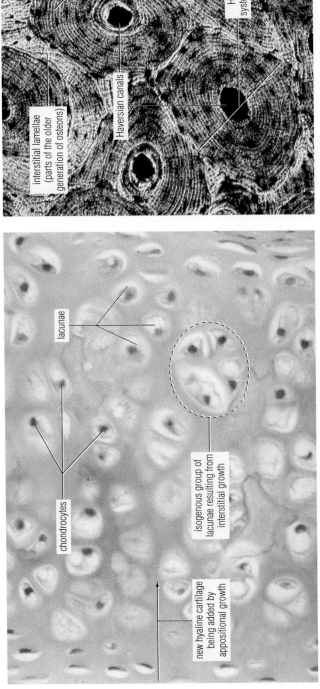

Figure 125a   **Hyaline cartilage** in a section of the trachea (prep. slide, 250×). (Photo by D. Morton)

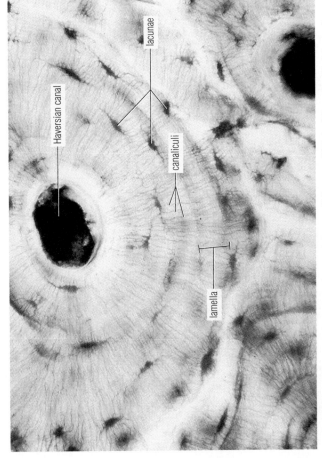

Figure 125c   **Haversian system** (osteon) (prep. slide, c.s., 500×). (Photo by D. Morton)

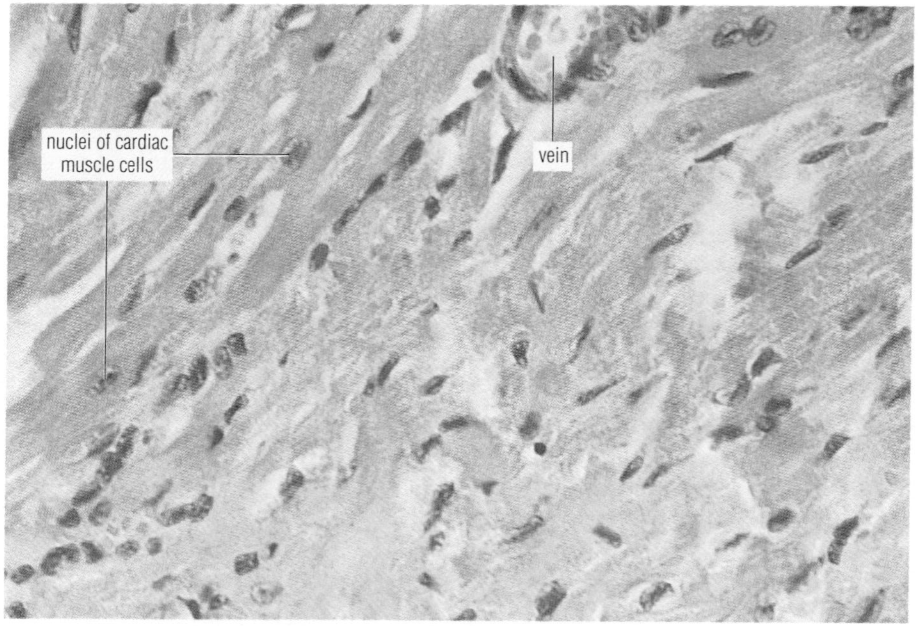

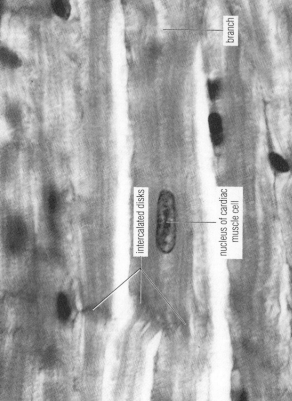

Figure 126a Cardiac muscle tissue in a section of the heart (prep. slide, 130×). (Photo by D. Morton)

nuclei of cardiac muscle cells

vein

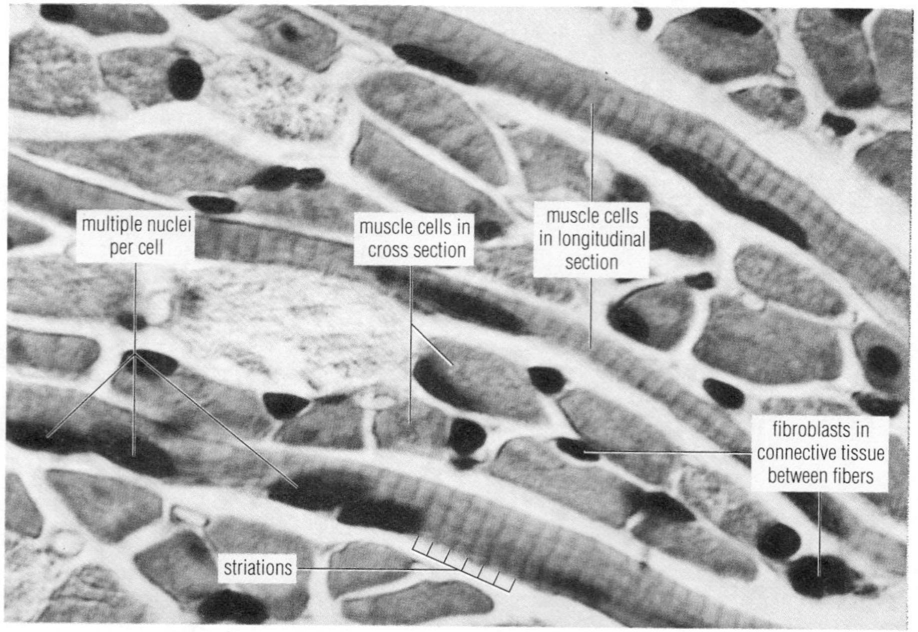

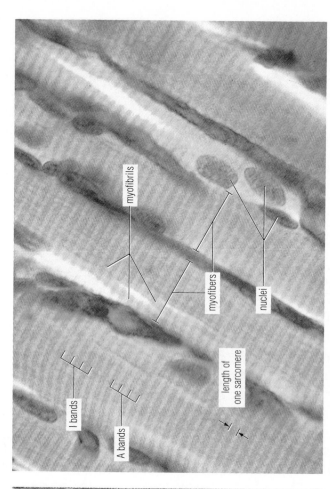

Figure 126c Skeletal muscle tissue in a section of the tongue (prep. slide, 500×). (Photo by D. Morton)

fibroblasts in connective tissue between fibers

muscle cells in longitudinal section

muscle cells in cross section

multiple nuclei per cell

striations

branch

intercalated disks

nucleus of cardiac muscle cell

Figure 126b Cardiac muscle cells. Striations are present but faint (prep. slide, l.s., 1200×). (Photo by D. Morton)

myofibrils

myofibers

nuclei

I bands

A bands

length of one sarcomere

Figure 126d Silver-stained skeletal muscle cells (prep. slide, l.s., 1200×). (Photo by D. Morton)

Figure 127b   **Smooth muscle cells** (prep. slide, l.s., 1300×). (Photo by D. Morton)

single smooth muscle cell with one central nucleus and without striations

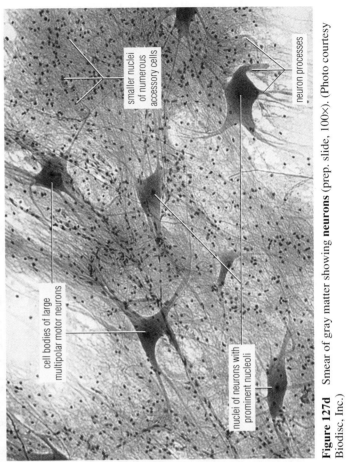

Figure 127d   Smear of gray matter showing **neurons** (prep. slide, 100×). (Photo courtesy Biodisc, Inc.)

smaller nuclei of numerous accessory cells

neuron processes

cell bodies of large multipolar motor neurons

nuclei of neurons with prominent nucleoli

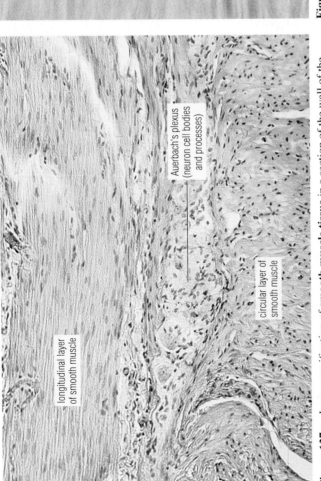

Figure 127a   Low magnification of **smooth muscle tissue** in a section of the wall of the small intestine (prep. slide, 150×). (Photo courtesy Biodisc, Inc.)

longitudinal layer of smooth muscle

Auerbach's plexus (neuron cell bodies and processes)

circular layer of smooth muscle

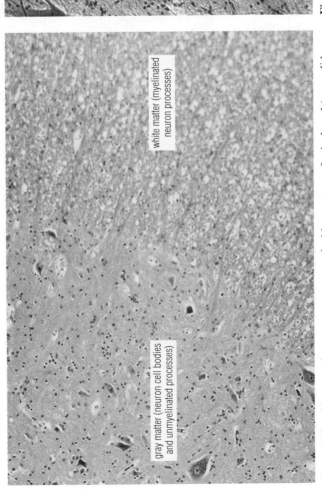

Figure 127c   **Nervous tissue**—gray matter and white matter—of spinal cord (prep. slide, c.s., 30×). (Photo by D. Morton)

white matter (myelinated neuron processes)

gray matter (neuron cell bodies and unmyelinated processes)

Figure 128a Hairy **skin** with **longitudinal sections of hair follicles** (prep. slide, 30×). (Photo by D. Morton)

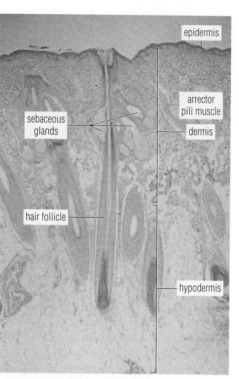

epidermis

arrector pili muscle

sebaceous glands

dermis

hair follicle

hypodermis

Figure 128b **Cross sections of hair follicles** at various levels (prep. slide, 80×). (Photo by D. Morton)

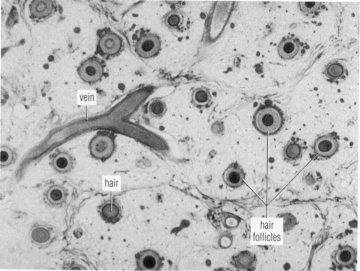

vein

hair

hair follicles

Figure 128c Section of thick skin showing **strata of epidermis** (prep. slide, 30×). (Photo by D. Morton)

stratum corneum

epidermis

stratum germinativum

stratum spinosum

stratum granulosum

stratum lucidum

dermis

Figure 128d **Pacinian** (pressure) **receptor** in a section of skin at the level of the hypodermis (prep. slide, 100×). (Photo by D. Morton)

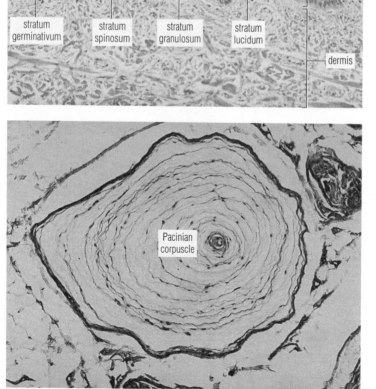

Pacinian corpuscle

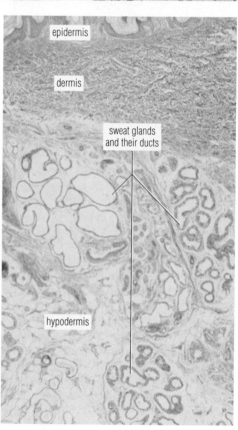

epidermis

dermis

sweat glands and their ducts

hypodermis

Figure 128e **Sweat glands** and their ducts in a section of the skin at the level of the hypodermis (prep. slide, 30×). (Photo by D. Morton)

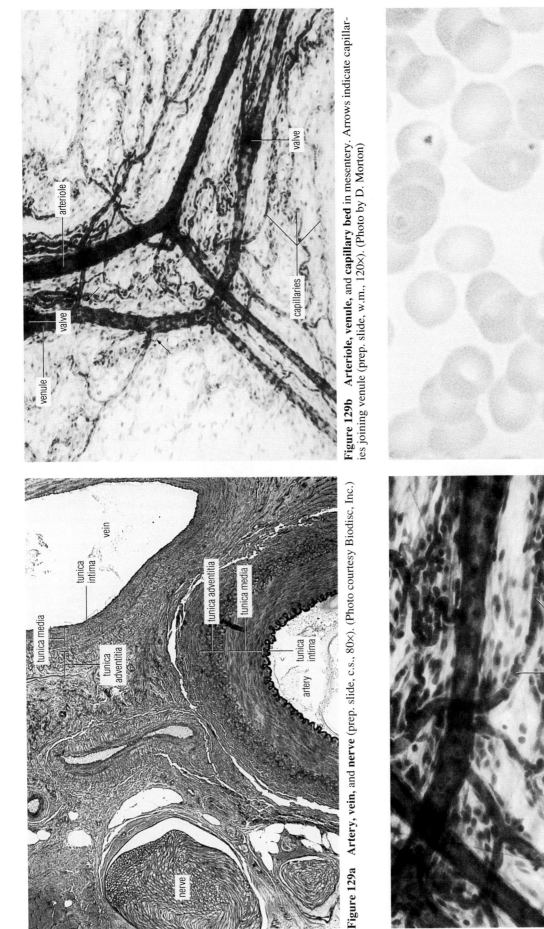

Figure 129b **Arteriole, venule,** and **capillary bed** in mesentery. Arrows indicate capillaries joining venule (prep. slide, w.m., 120×). (Photo by D. Morton)

Figure 129a **Artery, vein,** and **nerve** (prep. slide, c.s., 80×). (Photo courtesy Biodisc, Inc.)

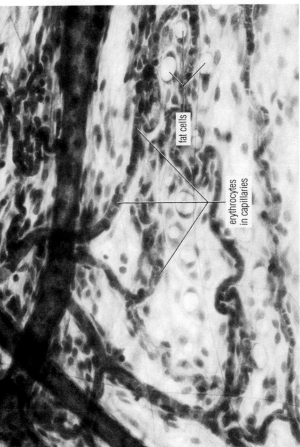

Figure 129d Smear of **human blood: erythrocytes** (prep. slide, w.m., 2000×). (Photo by D. Morton)

Figure 129c **Capillaries** (prep. slide, w.m., 500×). (Photo by D. Morton)

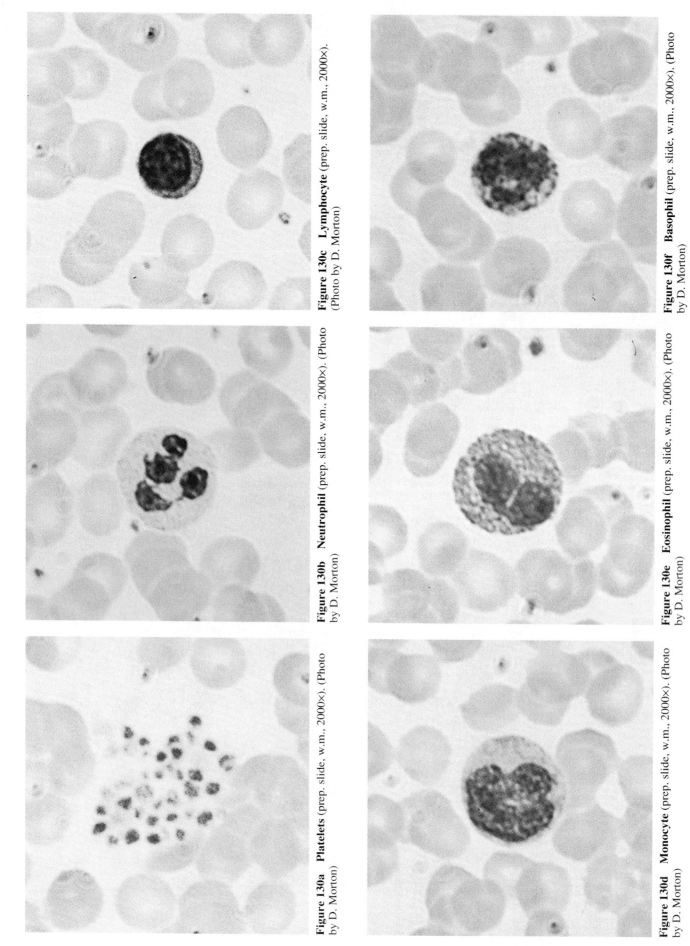

Figure 130c **Lymphocyte** (prep. slide, w.m., 2000×). (Photo by D. Morton)

Figure 130f **Basophil** (prep. slide, w.m., 2000×). (Photo by D. Morton)

Figure 130b **Neutrophil** (prep. slide, w.m., 2000×). (Photo by D. Morton)

Figure 130e **Eosinophil** (prep. slide, w.m., 2000×). (Photo by D. Morton)

Figure 130a **Platelets** (prep. slide, w.m., 2000×). (Photo by D. Morton)

Figure 130d **Monocyte** (prep. slide, w.m., 2000×). (Photo by D. Morton)

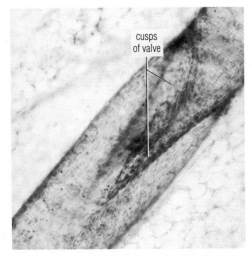

Figure 131a   Lymphatic vessel with **valve** (prep. slide, w.m., 90×). (Photo by D. Morton)

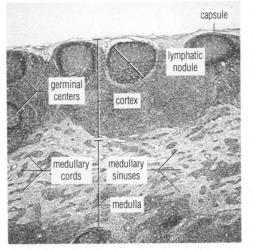

Figure 131b   Section of a **lymph node** (prep. slide, 20×). (Photo by D. Morton)

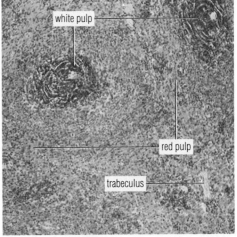

Figure 131c   Section of the **spleen** (prep. slide, 20×). (Photo by D. Morton)

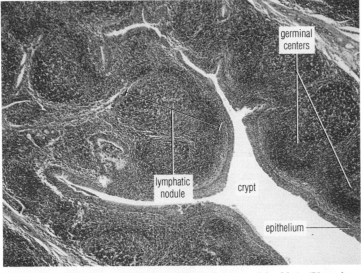

Figure 131d   Section of a **pharyngeal tonsil** (prep. slide, 20×). (Photo by D. Morton)

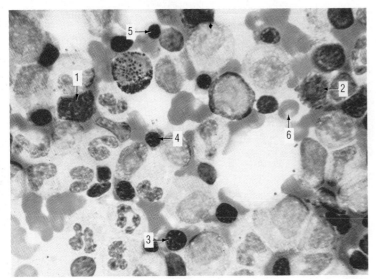

Figure 131e   Bone marrow smear. The numbered cells show the development of erythrocytes in chronological order, starting at the earliest stage present on the slide (prep. slide, w.m., 850×). (Photo by D. Morton)

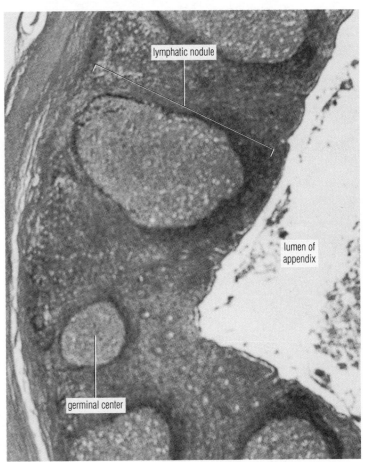

Figure 131f   Section of the **appendix** (prep. slide, 40×). (Photo by D. Morton)

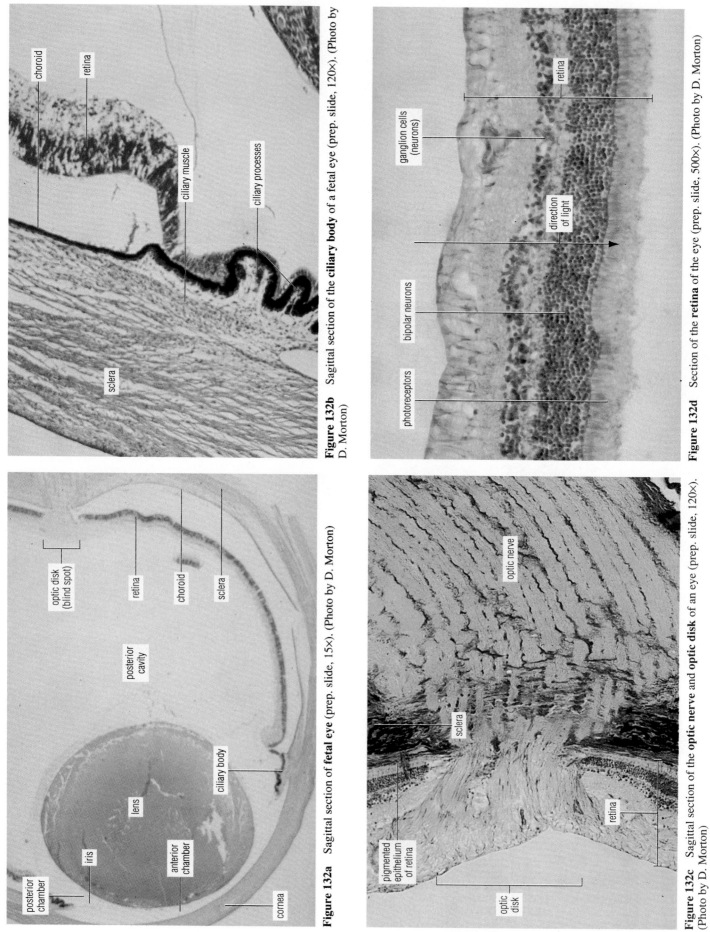

Figure 132a Sagittal section of **fetal eye** (prep. slide, 15×). (Photo by D. Morton)

Figure 132b Sagittal section of the **ciliary body** of a fetal eye (prep. slide, 120×). (Photo by D. Morton)

Figure 132c Sagittal section of the **optic nerve** and **optic disk** of an eye (prep. slide, 120×). (Photo by D. Morton)

Figure 132d Section of the **retina** of the eye (prep. slide, 500×). (Photo by D. Morton)

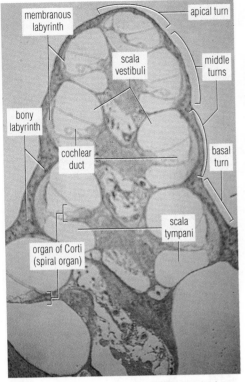

Figure 133a   Section of the **cochlea** of the inner ear (prep. slide, 25×). (Photo by D. Morton)

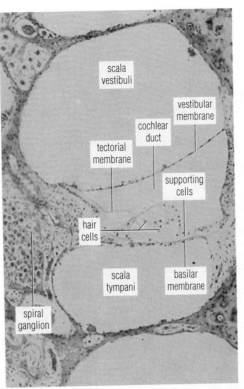

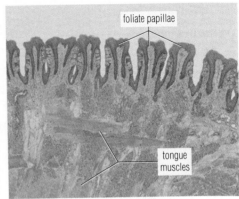

Figure 133c   Foliate papillae in a section of **rabbit tongue** (prep. slide, 25×). (Photo by D. Morton)

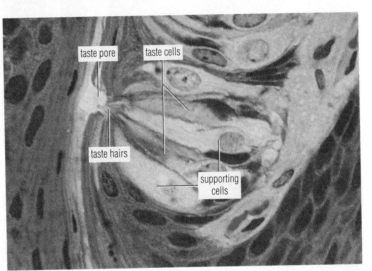

Figure 133e   Section of **taste bud** (prep. slide, 1000×). (Photo by D. Morton)

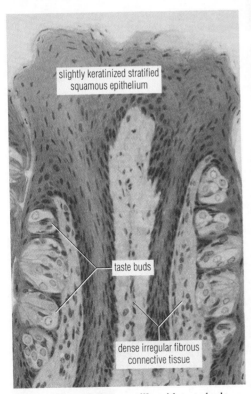

Figure 133d   **Foliate papilla** with taste buds (prep. slide, l.s., 250×). (Photo by D. Morton)

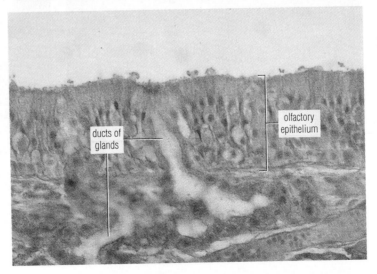

Figure 133f   Section of **olfactory epithelium** (prep. slide, 450×). (Photo by D. Morton)

Figure 134a
Anterior (left) **and**
posterior (right) **views**
of **human skeleton**.
The ilium is the most
superior of three bones
that fuse to form the os
coxa. The other two
bones are the pubis and
the ischium. (1/5×).
(Photos by D. Morton)

Figure 134b
Anterior (left) **and**
posterior (right) **views**
of **skull**. (2/3×).
(Photos by D. Morton)

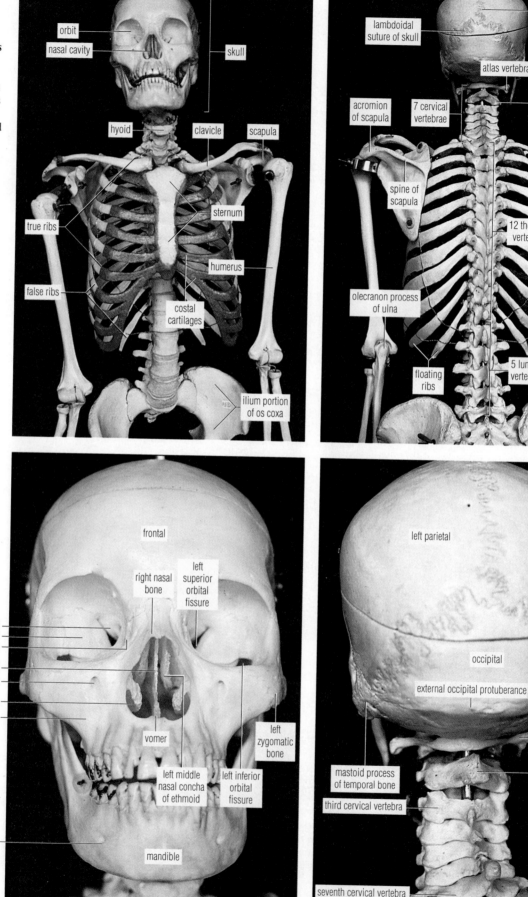

orbit
nasal cavity
skull
hyoid
clavicle
scapula
true ribs
sternum
false ribs
humerus
costal
cartilages
ilium portion
of os coxa

sagittal suture
of skull
lambdoidal
suture of skull
atlas vertebra
acromion
of scapula
7 cervical
vertebrae
axis vertebra
spine of
scapula
12 thoracic
vertebrae
olecranon process
of ulna
floating
ribs
5 lumbar
vertebrae

frontal
right nasal
bone
left
superior
orbital
fissure
right orbital plate of ethmoid
sphenoid
right lacrimal bone
perpendicular plate of ethmoid
right infraorbital foramen in maxilla
right inferior nasal concha
right maxilla
vomer
left
zygomatic
bone
left middle
nasal concha
of ethmoid
left inferior
orbital
fissure
right mental foramen in mandible
mandible

left parietal
occipital
external occipital protuberance
atlas vertebra
axis vertebra
mastoid process
of temporal bone
third cervical vertebra
seventh cervical vertebra

Figure 135a   Lateral view of skull. (1/3×). (Photo by D. Morton)

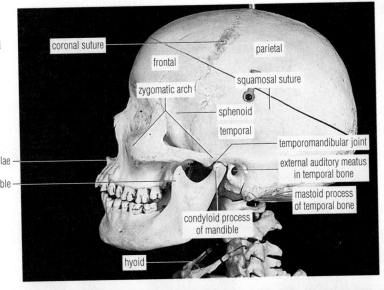

coronal suture

frontal

parietal

squamosal suture

zygomatic arch

sphenoid

temporal

temporomandibular joint

external auditory meatus in temporal bone

anterior nasal spine of maxillae

coronoid process of mandible

mastoid process of temporal bone

condyloid process of mandible

hyoid

Figure 135b
Types of **vertebrae**:
(1) atlas (first cervical vertebra)
(2) axis (second cervical vertebra)
(3) typical of the rest of the cervical vertebrae
(4) thoracic vertebra
(5) lumbar vertebra
(1/3×). (Photo by D. Morton)

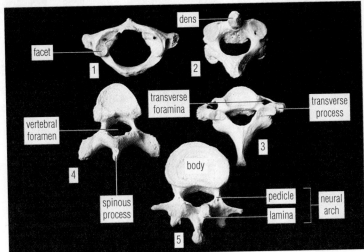

dens

facet

1

2

transverse foramina

transverse process

vertebral foramen

3

body

spinous process

pedicle

lamina

neural arch

4

5

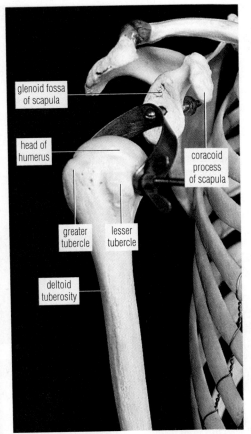

glenoid fossa of scapula

head of humerus

coracoid process of scapula

greater tubercle

lesser tubercle

deltoid tuberosity

Figure 135c   Anterior view of the **bones** of the **shoulder joint**. (2/3×). (Photo by D. Morton)

Figure 135d   Anterior view of the **bones** of the **wrist joint** and **hand**. The eight carpal bones are numbered:
(1) trapezium
(2) trapezoid
(3) capitate
(4) hamate
(5) triquetrum
(6) pisiform
(7) lunate
(8) scaphoid
(1/2×). (Photo by D. Morton)

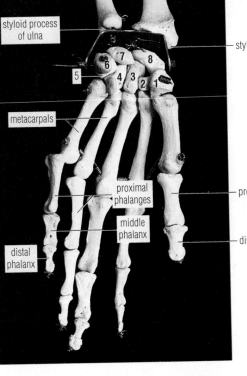

styloid process of ulna

styloid process of radius

7   8

5

6

4 3 2 1

metacarpals

proximal phalanges

proximal phalanx

middle phalanx

distal phalanx

distal phalanx

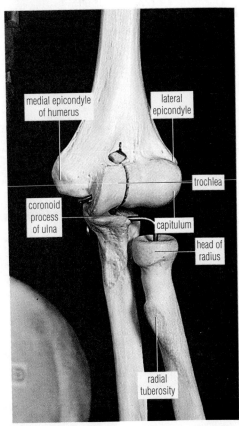

medial epicondyle of humerus

lateral epicondyle

trochlea

coronoid process of ulna

capitulum

head of radius

radial tuberosity

Figure 135e   Anterior view of the **bones** of the **elbow joint**. (3/4×). (Photo by D. Morton)

Figure 136a Anterior view of the bones of the female pelvis and hip joint. Compared to that of the female, the male pelvis is relatively heavier with more prominent processes, its inlet is more heart-shaped, and its outlet is narrower. Also, the pubic arch of the male pelvis is more acute (less than 90°) and the tip of the coccyx tilts more anteriorly than in the female. (1/2×). (Photo by D. Morton)

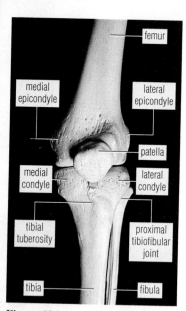

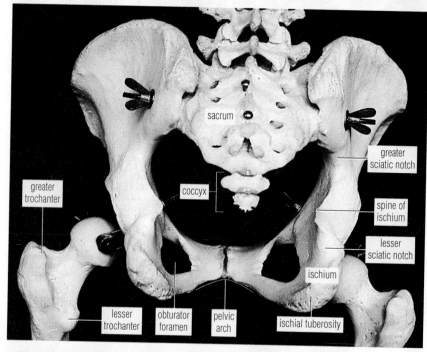

Figure 136b Posterior view of the bones of the female pelvis and hip joint. (1/2×). (Photo by D. Morton)

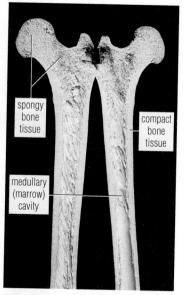

Figure 136c Anterior view of the bones of the knee joint. (1/4×). (Photo by D. Morton)

Figure 136d Lateral view of the bones of the ankle joint and foot. The seven tarsal bones are numbered:
(1) talus
(2) first cuneiform
(3) second cuneiform
(4) third cuneiform
(5) cuboid
(6) naviculus
(7) calcaneus
(2/5×). (Photo by D. Morton)

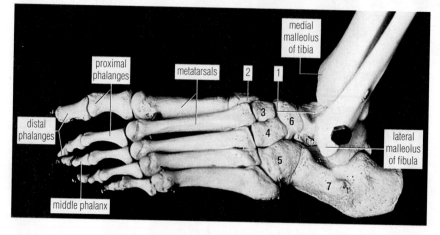

Figure 136e Sawed femur. (1/4×). (Photo by D. Morton)

**Figure 137a
External anatomy** of
the **sheep eye** (2/3×).
(Photo by D. Morton)

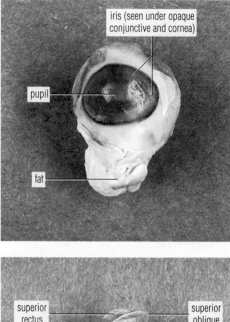

iris (seen under opaque
conjunctive and cornea)

pupil

fat

**Figure 137b
External** (extrinsic)
eye muscles of sheep
(2/3×). (Photo by
D. Morton)

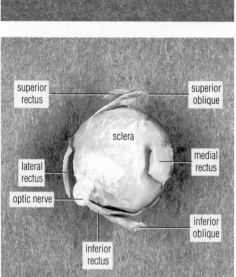

superior
rectus

superior
oblique

sclera

lateral
rectus

medial
rectus

optic nerve

inferior
oblique

inferior
rectus

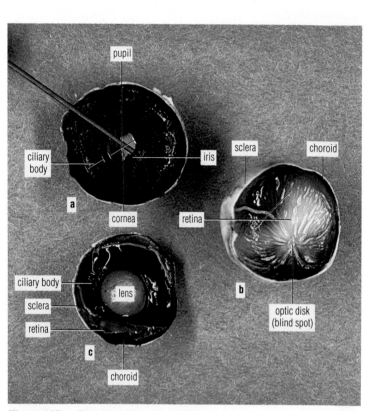

pupil

ciliary
body

iris

sclera

choroid

a

cornea

retina

ciliary body

sclera

retina

lens

b

c

choroid

optic disk
(blind spot)

Figure 137c   Dissected sheep eye: (a) inside of front of eyeball with lens
removed; **(b)** inside of back of eyeball; **(c)** inside of front of eyeball (1×).
(Photo by D. Morton)

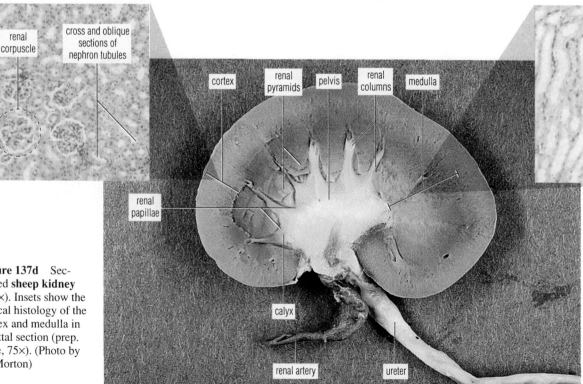

renal
corpuscle

cross and oblique
sections of
nephron tubules

cortex

renal
pyramids

pelvis

renal
columns

medulla

collecting
ducts

vasa recta
(blood vessels)

renal
papillae

calyx

renal artery

ureter

Figure 137d   Sec-
tioned **sheep kidney**
(2/3×). Insets show the
typical histology of the
cortex and medulla in
sagittal section (prep.
slide, 75×). (Photo by
D. Morton)

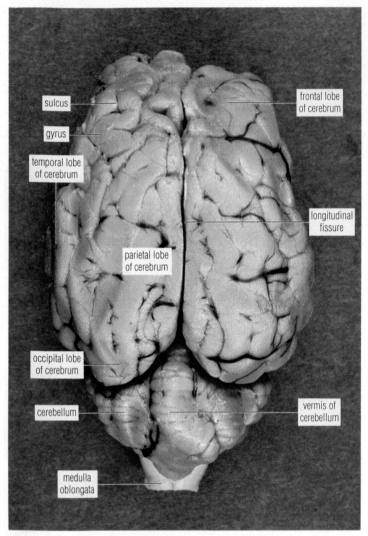

Figure 138a **Dorsal view** of **external anatomy** of **sheep brain** (1×). (Photo by D. Morton)

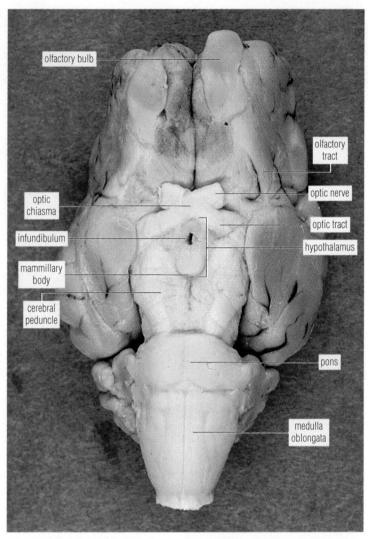

Figure 138b **Ventral view** of **external anatomy** of **sheep brain** (1×). (Photo by D. Morton)

Figure 138c Longitudinally **sectioned sheep brain** (1×). (Photo by D. Morton)

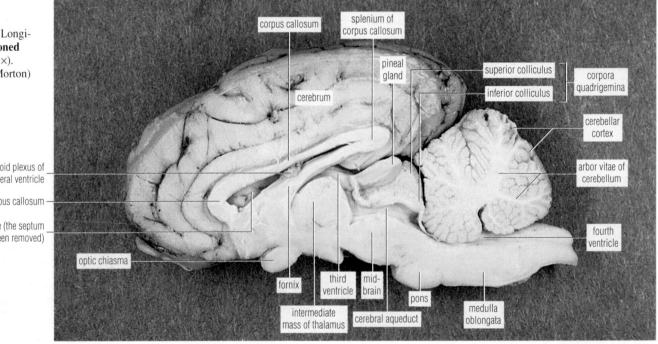

Figure 139a   Ventral (left) **and dorsal** (right) **views** of **sheep heart** with **dissected vessels** (2/3×). (Photos by D. Morton)

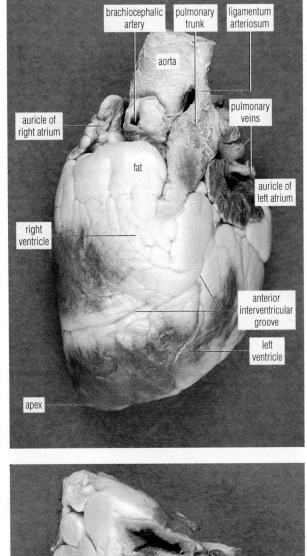

brachiocephalic artery

pulmonary trunk

ligamentum arteriosum

aorta

auricle of right atrium

pulmonary veins

fat

auricle of left atrium

right ventricle

anterior interventricular groove

left ventricle

apex

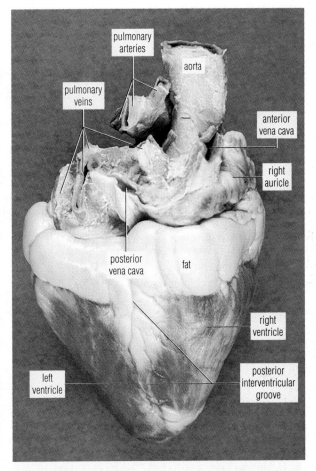

pulmonary arteries

aorta

pulmonary veins

anterior vena cava

right auricle

posterior vena cava

fat

right ventricle

left ventricle

posterior interventricular groove

Figure 139b Ventral (left) **and dorsal** (right) **views** of **sectioned sheep heart** (2/3×). (Photos by D. Morton)

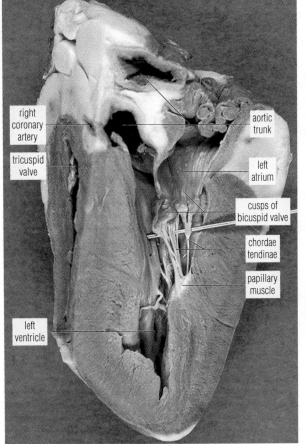

right coronary artery

aortic trunk

tricuspid valve

left atrium

cusps of bicuspid valve

chordae tendinae

papillary muscle

left ventricle

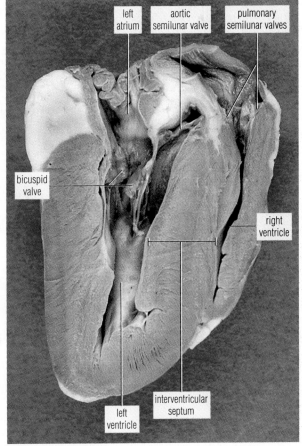

left atrium

aortic semilunar valve

pulmonary semilunar valves

bicuspid valve

right ventricle

interventricular septum

left ventricle

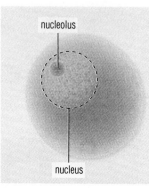

Figure 140a   Sea star development: zygote (prep. slide, w.m., 250×). (Photo by D. Morton)

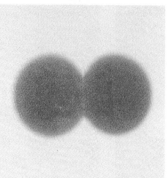

Figure 140b   Two-cell stage (prep. slide, w.m., 250×). (Photo by D. Morton)

Figure 140c   Four-cell stage (prep. slide, w.m., 250×). (Photo by D. Morton)

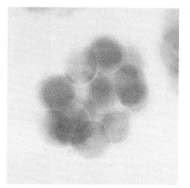

Figure 140d   Early morula (prep. slide, w.m., 250×). (Photo by D. Morton)

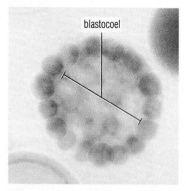

Figure 140e   Early blastula (prep. slide, w.m., 250×). (Photo by D. Morton)

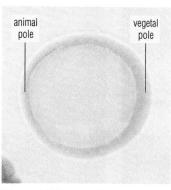

Figure 140f   Late blastula (prep. slide, w.m., 250×). (Photo by D. Morton)

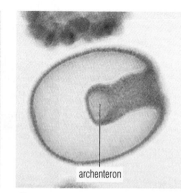

Figure 140g   Early gastrula (prep. slide, w.m., 250×). (Photo by D. Morton)

Figure 140h   Mid gastrula (prep. slide, w.m., 200×). (Photo by D. Morton)

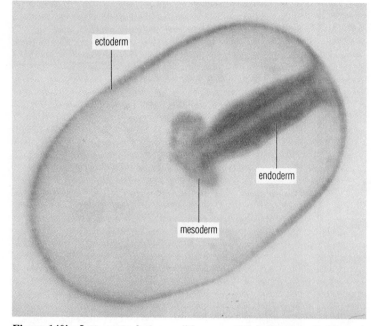

Figure 140i   Late gastrula (prep. slide, w.m., 300×). (Photo by D. Morton)

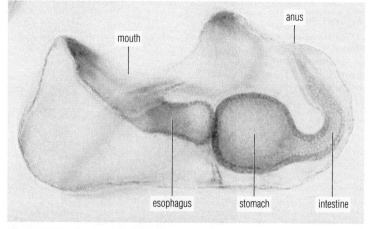

Figure 140j   Bipinnaria larva (prep. slide, w.m., 250×). (Photo by D. Morton)

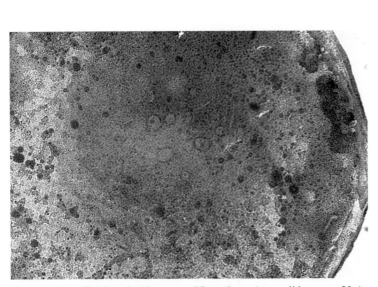

Figure 141a   Chicken development: blastoderm (prep. slide, w.m., 30×). (Photo courtesy Biodisc, Inc.)

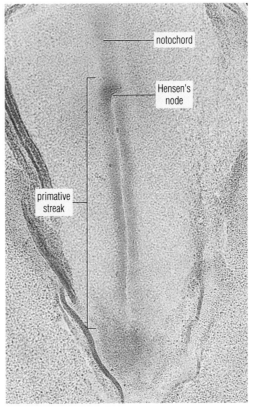

Figure 141b   Gastrulation (18 hours of incubation) (prep. slide, w.m., 40×). (Photo courtesy Biodisc, Inc.)

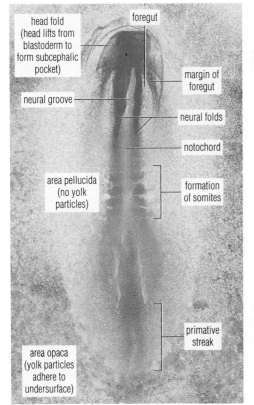

Figure 141c   Embryo of **21 hours** of incubation (prep. slide, w.m., 20×). (Photo courtesy Biodisc, Inc.)

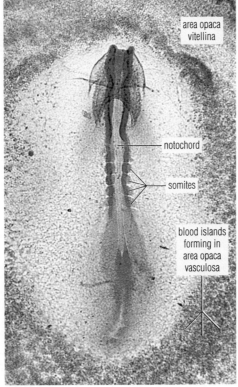

Figure 141d   Embryo of **24 hours** of incubation (prep. slide, w.m., 20×). (Photo courtesy Biodisc, Inc.)

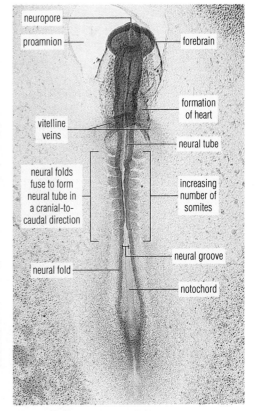

Figure 141e   Embryo of **28 hours** of incubation (prep. slide, w.m., 20×). (Photo courtesy Biodisc, Inc.)

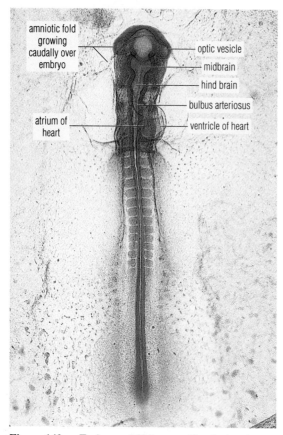

Figure 142a   Embryo of **38 hours** of incubation (prep. slide, w.m., 20×). (Photo courtesy Biodisc, Inc.)

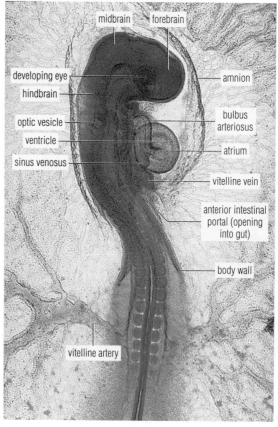

Figure 142b   Embryo of **44 hours** of incubation (prep. slide, w.m., 16×). (Photo courtesy Biodisc, Inc.)

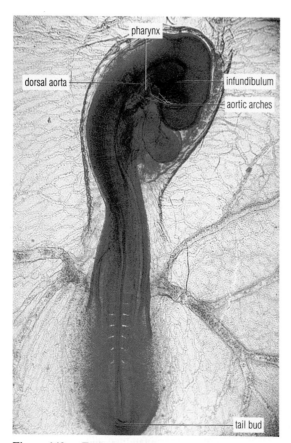

Figure 142c   Embryo of **48 hours** of incubation (prep. slide, w.m., 13×). (Photo courtesy Biodisc, Inc.)

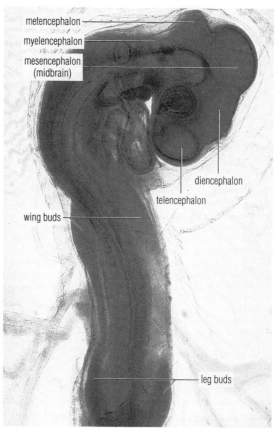

Figure 142d   Embryo of **72 hours** of incubation (prep. slide, w.m., 13×). (Photo courtesy Biodisc, Inc.)

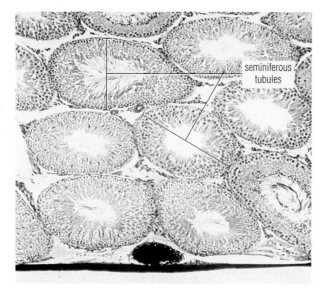

Figure 143a   Section of **mammalian testis** (prep. slide, 90×). (Photo courtesy Biodisc, Inc.)

seminiferous tubules

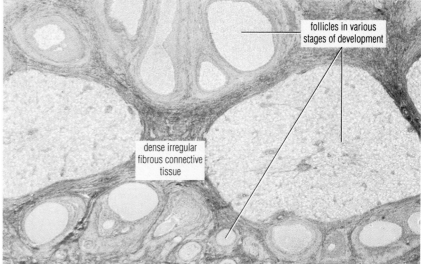

follicles in various stages of development

dense irregular fibrous connective tissue

Figure 143b   Section of **mammalian ovary** (prep. slide, 30×). (Photo by D. Morton)

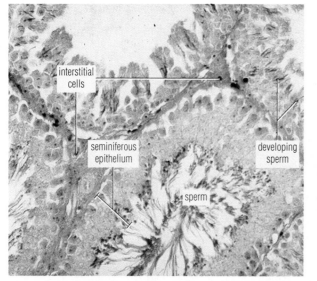

interstitial cells

seminiferous epithelium

developing sperm

sperm

Figure 143c   **Seminiferous tubules** (prep. slide, c.s., 400×). (Photo by D. Morton)

primary oocytes in primary follicles

Figure 143d   **Primary follicles** (prep. slide, sec., 450×). (Photo by D. Morton)

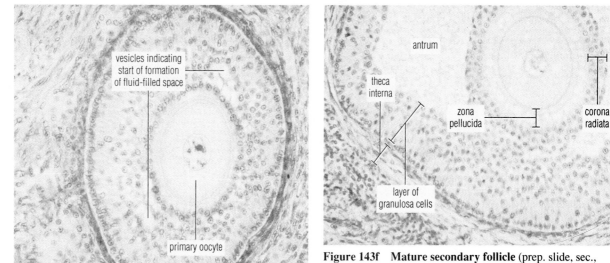

vesicles indicating start of formation of fluid-filled space

primary oocyte

Figure 143e   **Maturing secondary follicle** (prep. slide, sec., 400×). (Photo by D. Morton)

antrum

theca interna

layer of granulosa cells

zona pellucida

corona radiata

Figure 143f   **Mature secondary follicle** (prep. slide, sec., 350×). (Photo by D. Morton)

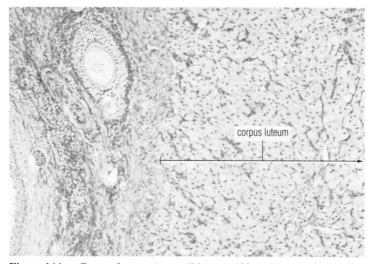

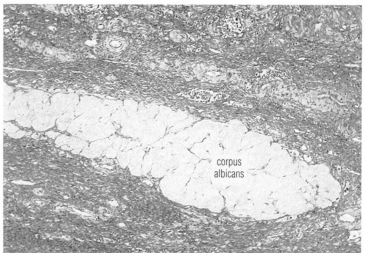

Figure 144a Corpus luteum (prep. slide, sec., 100×). (Photo by D. Morton)

Figure 144b Corpus albicans (prep. slide, sec., 100×). (Photo courtesy Biodisc, Inc.)

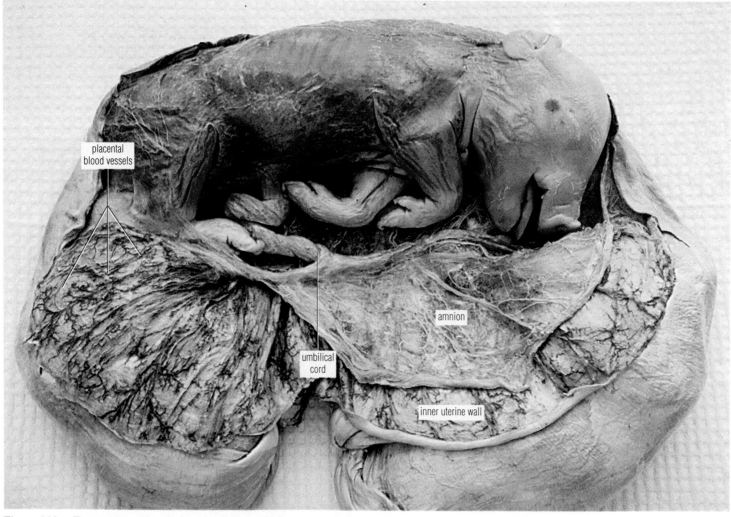

Figure 144c Extraembryonic membranes of the **fetal pig**. Fetal vessels have been injected with yellow latex. Maternal vessels have been injected with red latex for arteries and blue latex for veins. (3/4×). (Specimen courtesy Wards Natural Science Establishment, Inc.; Photo by D. Morton)